A NATURALIST'S GUIDE TO THE

GARDEN FLOWERS
OF
INDIA
Pakistan, Nepal, Bhutan, Bangladesh and Sri Lanka

Pradeep Sachdeva and Vidya Tongbram

JOHN BEAUFOY PUBLISHING

First published in the United Kingdom in 2019 by John Beaufoy Publishing Ltd
11 Blenheim Court, 316 Woodstock Road, Oxford OX2 7NS, U.K.
www.johnbeaufoy.com

Photo Credits
Front cover Clockwise from top Queen's Crape Myrtle, Chalice Vine, Lotus, White Ginger Lily
(Pradeep Sachdeva and Vidya Tongbram)
Back cover Calendula (Pradeep Sachdeva)
Title page Lotus (Vidya Tongbram)
Contents page Rain Lily (Vidya Tongbram)

All photographs by Pradeep Sachdeva and Vidya Tongbram, except Oken Tayeng 10, 88b, 89, Arti Mathur 170,
Gautam Sachdeva 105, 110t, WikiCommons 5,12m, 13t, 47b, 124b

ISBN 978-1-912081-75-2

Edited by Krystyna Mayer
Designed by Gulmohur Press, New Delhi
Printed and bound in Malaysia by Times Offset (M) Sdn. Bhd.

·CONTENTS·

INTRODUCTION

Flowers are everywhere in the Indian subcontinent. They are common and they are special. They are found in gardens, and used in ceremonial rituals, herbal remedies, perfumes and cosmetics. They feature in literature, works of art and crafts. Flowers are tradition. They are for daily prayers, weddings, births and last rites, and everything in between.

Hindus associate different flowers with various gods and goddesses. Laxmi, goddess of wealth, is usually depicted sitting on a pink lotus and holding two more lotuses. Saraswati, goddess of knowledge, is depicted sitting on a white lotus. Its roots submerged in mud, the lotus emerges high above the murky water to blossom radiantly. It symbolizes divinity, purity, transcendence and enlightenment. It is revered by both Hindus and Buddhists. This reverence is evident in its representation in religious iconography. The lotus *Nelumbo nucifera* is the national flower of India.

On festive occasions, Hindus decorate their floors with floral carpet patterns known as Rangoli. Some of the most beautiful Rangoli are made in the southern state of Kerala during Onam, its harvest festival. They are made by the women of the house in the front courtyard. During Diwali, the Hindu Festival of Light, every household is decorated with strings of marigolds. The Sikhs shower their holy book, the *Guru Granth Sahib*, with flowers. Marriages are consecrated by the bride and groom exchanging garlands. In eastern parts of India, a bride strings Kundo, or jasmine flowers, to create a garland in a ceremony of enormous significance. Women adorn their hair with fragrant blooms. They tuck Champa (Michelia) flowers in their buns, or trail strings of Mogra (jasmine) along plaited braids to perfume the hair. In South India, it is most common to see women coiffure their long, oiled braids with strings of scented white jasmine. Elaborate versions are worn by the bride on her wedding day, and by Indian classical dancers in their performances.

GARDEN HERITAGE & FLOWERS IN EARLY TEXTS

In ancient India, flowers were given an important place in the art of decorating male and female bodies alike. Numerous ancient Indian texts such as the *Vedas*, the *Upanishads* and the *Puranas*, to name a few, mention flowers and plants in general and their importance in life. The earliest of these texts, the *Rigveda*, dates back to 1800 BC. The texts have elaborate descriptions of the botanical classifications of plants, and the uses of the plants, as well as their methods of propagation. The Indian sages who wrote the texts had expertise in the uses of different flowers for therapeutic, religious and general uses.

Ancient Indian literature also has numerous texts devoted exclusively to flowers. Most remain in manuscript form in repositories across India. They include *Pushpachintamani*, *Pushpa Maahaatmyam*, *Pushparatnakara Tantram*, *Parijatamanjari* and *Pushpayurveda*. The last text is ascribed to the Jain monk Samantbhadra Svami, who lived during the latter part of the second century BC. and named 18,000 flowers.

India's garden heritage speaks of temple gardens, sacred forests, palace gardens, Mughal gardens and colonial British gardens. Most of our knowledge about temple and palace gardens is from references in ancient literature and mythology. Little material evidence remains of these gardens, but one such garden of lore that still exists today is Nidhi Van

Goddess Laxmi associated with the sacred lotus, painting by Raja Ravi Verma

in Vrindavan, Uttar Pradesh. It is a grove of small, entwining trees known locally as *Vana Tulsi*. The garden is considered sacred and venerated even today: it is where the Divine Play of Lord Krishna and his *Gopis* is believed to have taken place.

THE MUGHALS & FLOWERS

The Mughals ruled India for about 300 years, from AD 1526. Their empire was the third largest empire in the Indian subcontinent, after the Mauryan and British empires. In the seventeenth century, India under the Mughals had become the world's leading economic power. The Mughals had a love of beauty, and some of the finest monuments in south Asia were built in the Mughal Era. These include the Red Fort and Jama Masjid in Delhi, and the Fort and Taj Mahal in Agra.

The Mughals' pursuit of beauty was also expressed in the numerous gardens they built all over their empire, in Kabul, Lahore, Srinagar, Agra, Jaipur, Delhi, Bharatpur, Mandu and Kalka. The Mughal gardens of Kashmir – Shalimar Bagh, Nishat Bagh and Chashme-Shahi – are some of its best living treasures.

To the Mughals also goes the credit for introducing the much-cherished tulips. Their passion for plants, including flowers, is well documented in memoirs, paintings, and the use of floral motifs in inlays, carvings and murals in buildings. The memoirs of Babur and Jahangir are filled with details of their gardens. Monument Gardens such as those at Humayun's Tomb and Sunder Nursery in Delhi are today being restored to a glory that matches the contemporary aesthetics of Mughal gardens.

Floral motifs in Mughal architecture

Flowers & the British in India

The British, in their 300-year rule in the subcontinent, brought a sea change in the garden and flower culture of India. While building new cities and redeveloping old ones, the British added to the landscape of Indian gardens by introducing vast lawns lined with beds of flowering annuals and perennials that grew in the British Isles and continental Europe. The gardens were of all shapes and sizes, and were public as well as private.

The British brought numerous flowering plants to the subcontinent, from both England and the colonies. They built many botanical gardens across India. One major repository of plants was the Agri Horticultural Society of India in Kolkata. Established in 1820, it has greenhouses, flower gardens, a laboratory, a library and a significant collection of flowering plants. Some other significant gardens that still exist are the Indian Botanic Garden in Kolkata, Cubbon Park and Lal Bagh in Bengaluru, Lloyd's Botanic Garden in Darjeeling, and the gardens in Saharanpur, Ootacamund and Perediniya in Kandy, Sri Lanka.

For centuries, plant hunters and collectors have travelled the world and have discovered, cultivated, hybridized and transported plants from one part of the globe to another. The Indian subcontinent has also attracted its fair share of plant hunters for its unique plants. One of them was Joseph Hooker, who made an expedition to Sikkim in 1848–1849. The Himalayan Blue Poppy, rhododendrons, *Primula sikkimensis*, some species of clematis, and species of magnolia thus travelled from the Himalayas across the world. Another plant hunter who operated in the Himalayas was Frank Kingdon Ward.

Lal Bagh in Bangalore, established by the Sultans and expanded by the British

PLANT HABITATS OF THE INDIAN SUBCONTINENT

There is an extraordinary variety of flowers in the subcontinent. The region is estimated to have more than 18,000 species of flowering plant growing in different ecosystems. These range from the hot and arid desert habitats of Rajasthan, to the fertile areas of the Indo-Gangetic plains; from the conifers and broadleaved plants of the eastern and western Himalayas, to the dry scrub of the Punjab; from the tropical green rainforests of North-east India to Kerala in the south. Then there is the distinct landscape of the Andaman and Nicobar Islands in the Bay of Bengal, similar to that of Southeast Asia. Many plants are native to the subcontinent, and many have been introduced over millennia by merchants and colonizers, including the Arabs, Portuguese, British and Mughals. Different flower and plant habitats in the country can be classified into separate regions.

Pakistan has a wide variety of zones ranging from permanent snowfields, subtropical pine, subtropical dry mixed scrubland as well as sand dune desert. Sri Lanka has an eco-system closer to the Western Ghats in large parts while Bangladesh which is on the delta of the Ganges has moist deciduous forests and mangroves.

Trans Himalayas North of the main Himalayan Range, covering Ladakh and the Tibetan plateaus. Cold and arid climate conditions allow sparse alpine vegetation.
Western Himalayas Evergreen and deciduous broadleaved forests stretching from Kashmir, to Kumaon, to west Nepal.
Eastern Himalayas Temperate subalpine coniferous forests, subtropical jungle, savannah and rich alpine meadows stretching from North Bengal, to Sikkim, to Arunachal Pradesh.
North-west Desert Region Thar Desert, stretching from Punjab, Haryana and Rajasthan, to Gujarat.
Semi-Arid Region Arid desert regions of Thar and parts of Rajasthan. Natural vegetation comprises thorny, dry deciduous forests.
Gangetic Plains Stretching from the plains of west Uttar Pradesh to the Hooghly Delta and the forests of Sundarbans.
Western Ghats Stretching from South Gujarat to Kanyakumari, this region features four distinct eco-regions – the North Western Ghats moist deciduous forests, the North Western Ghats rainforests, the southwestern Ghats moist deciduous forests and the South Western Ghats rainforests.
Deccan Peninsula The Satpura Range, tropical thorn forests, and the mangroves on the delta of the Godavari River. In Peninsular India south of the Godavari River, cutting across Maharashtra, Andhra Pradesh, Karnataka and Tamil Nadu, the Central Deccan Plateau dry deciduous forests.
North-east India Evergreen rainforests, tropical semi-evergreen forests and subtropical hill forests.
Andaman and Nicobar Islands Mangroves and wetlands, and evergreen, semi-evergreen and deciduous forests.
Coasts West and East coastal belt, whose vegetation comprises fresh and brackish wetlands, and mangroves.

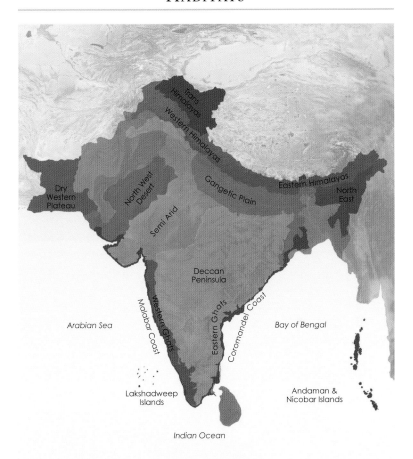

The climate of the subcontinent is governed by the monsoon regime, with rain concentrated in a small period of the year. The major seasons are the monsoon (June–September), post-monsoon (October–November), winter (December–February) and summer (March–May). There are no defined 'hardiness zones' for south Asia. The different climatic sub-types define the flower types grown in the areas.

Flora Hotspots in the Region

Rhododendron forests in Nagagigi, Arunachal Pradesh

The Indian subcontinent is home to some unique meadows and valleys, where many endemic varieties of flora are found. The best known is the Valley of Flowers in the Western Himalayas. Located at an altitude of 3,350-3,670m above sea level, it has a wide variety of alpine flowers. Some other locations are the Kaas Plateau in the Western Ghats, with about 850 flower varieties; the Yumthang Valley of Flowers in the Sikkim Himalayas; Dzukou Valley in Manipur and Nagaland in Northeast India; and the Nagagigi in the West Kamang area of Arunachal Pradesh, which has a wide variety of rhododendrons.

The legacy of garden flowers in India is therefore an amalgamation of native, introduced and naturalized, and collected exotic plants that have been inherited over India's long and colourful history. This book is a field guide for gardeners and flower enthusiasts alike. It provides descriptions and photographs for easy identification of flowers, and outlines their potential in the garden and their other uses, both medicinal and ornamental.

The Significance of Flowers & Their Parts

The flowers play an essential part in the survival of a plant species. The dazzling colours, shapes and smells that draw people to flowers are actually present to attract pollinators that enable the plants to reproduce. Flowering plants, or angiosperms, have an ability to evolve and adapt the process of their regeneration and reproduction. Flowers by definition are 'seed-bearing parts of a plant consisting of sexual organs that are typically surrounded by brightly coloured petals'.

As stated by naturalist David Attenborough, the 'flower is the reproductive structure of flowering plants'.

To reproduce, flowering plants depend on wind or insects and other animals for pollen dispersion. Wind dispersion is a simple phenomenon in which a gust of breeze carries pollen to its female receptors to fertilize the ovules. Dispersion by an external pollinator, such as a bee or a bird, is a reciprocal process that demonstrates the co-dependency between a flower and its pollinator.

Nature exhibits intelligence in its selection of the design and appearance of a flower, depending on the means by which a plant proliferates. Plants that depend on wind dispersion of pollen – such as grasses – have inconspicuous flowers, while those that depend on external pollinators have designer flowers. Their shape, size, colour, smell and nectar are all uniquely determined and coordinated to draw the attention of specific pollinators, be it bees, birds, bats, moths or butterflies.

Oriental Lily

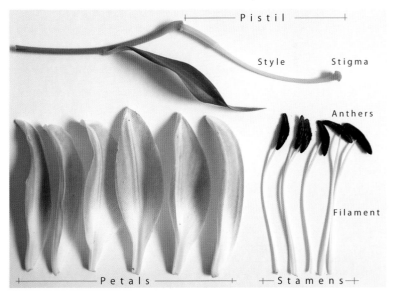

Pistil

Style   Stigma

Anthers

Filament

Petals   Stamens

Components of an Oriental Lily flower

Saucer shaped

Flat faced or salverform

Tube shaped or tubular

Urn shaped or urceolate

Star shaped or stellate

Cross shaped or cruciform

Bell shaped or campanulate

Trumpet shaped

Funnel shaped

Cup shaped or cupulate

Pea shaped

Two lipped or labiate

CLASSIFICATION & FAMILIES

Plants in this book have been categorized in alphabetical order according to the common name of the family. Swedish naturalist Carolus Linnaeus classified plants in family groups according to the arrangement of the reproductive components in the flowers. In a family group, plants often share a similarity in the shape and arrangement of the flowers.

No.	Group	Family Name
1	Acanthuses	Acanthaceae
2	Achiotes	Bixaceae
3	Agaves	Agavaceae
4	Aloes	Asphodelaceae
5	Amaryllises	Amaryllidaceae
6	Arrowroots	Marantaceae
7	Arums	Araceae
8	Balsams	Balsaminaceae
9	Bananas	Musaceae
10	Barberries	Berberidaceae
11	Barbados Cherries	Malpighiaceae
12	Begonias	Begoniaceae
13	Bignonias	Bignoniaceae
14	Bindweeds	Convolvulaceae
15	Birthworts	Aristolochiaceae
16	Borages	Boraginaceae
17	Cannonball Tree	Lecythidaceae
18	Buttercups	Ranunculaceae
19	Buckwheats	Polygalaceae
20	Capers	Capparaceae
21	Camellias	Theaceae
22	Cannas	Cannaceae
23	Carnations	Caryophyllaceae
24	Coffees	Rubiaceae
25	Costuses	Costaceae
26	Combretales	Combretaceae
27	Daisies	Asteraceae
28	Dogbanes	Apocynaceae
29	Evening Primroses	Onagraceae
30	Figworts	Scrophulariaceae
31	Four O'Clocks	Nyctaginaceae
32	Geraniums	Geraniaceae
33	Gingers	Zingiberaceae
34	Heathers	Ericaceae
35	Heliconias	Heliconiaceae

No.	Group	Family Name
36	Honeysuckles	Caprifoliaceae
37	Hydrangeas	Hydrangeaceae
38	Ironwood	Calophyllaceae
39	Irises	Iridaceae
40	Leadworts	Plumbaginaceae
41	Legumes	Fabaceae
42	Lillies	Liliaceae
43	Loosestrifes	Lythracea
44	Lotuses	Nelumbonaceae
45	Magnolias	Magnoliaceae
46	Mallows	Malvaceae
47	Melastomes	Melastomaceae
48	Mints	Lamiaceae
49	Myrtles	Myrtaceae
50	Nasturtiums	Tropaeolaceae
51	Nightshades	Solanaceae
52	Ochnas	Ochnaceae
53	Olives	Oleaceae
54	Orchids	Orchidaceae
55	Passion Flowers	Passifloraceae
56	Peony	Paeoniaceae
57	Pickerelweeds	Pontederiaceae
58	Poppies	Papaveraceae
59	Primroses	Primulaceae
60	Roses	Rosaceae
61	Soursops	Annonaceae
62	Stonecrops	Crassulaceae
63	Strelitzias	Strelitziaceae
64	Verbenas	Verbenaceae
65	Violets	Violaceae
66	Water Plantains	Alismataceae
67	Water Lillies	Nymphaeaceae
68	Willows	Salicaceae
69	Wood Sorrels	Willows

Glossary

accrescent Growing larger after flower maturity, most often applied to calyx.

actinomorphic Usually applied to flowers. Having multiple planes so that any line drawn through the middle produces two mirror image halves.

alternipetalous Of floral parts whose positions in flowers alternate with those of petals.

annual plant Refers to a plant whose life cycle is one year.

anther Pollen-bearing part of a stamen.

beard Clump of hairs that appears in some irises from the central part.

bicoloured Having two colours, usually in flowers.

biennial plant Plant whose life cycle occurs in two years.

bloom Flower or inflorescence.

blossom Flower or inflorescence.

bract Leaf-like leaf part that is sometimes showy and brightly coloured, located below a flower, flower stalk or inflorescence.

bud Immature flower, leaf or stem still enclosed by its protective covering.

calyx Collective term for sepals that protect flower bud.

campanulate Bell shaped.

carpel Female reproductive organs of flowering plants comprising an ovary, often a style and a stigma. Carpels are separate or fused to form a single pistil.

catkin Dense, cylindrical, often drooping cluster of unisexual inflorescence.

corolla Petals of a flower considered as a group.

cross pollination Process by which pollen from one plant lands and germinates on another plant.

cultigen Plant known only in cultivation without any known examples in the wild.

cultivar Plant that has been created and maintained through cultivation.

filament Stalk bearing anther in a stamen.

floret Small flowers or single flower within an inflorescence.

hardiness Ability of a plant to survive in certain conditions or locations.

hybrid Plant produced via sexual reproduction beween two different species or varieties.

inflorescence Branched or unbranched stalks on which flowers are arranged in a characteristic way.

invasive Generally refers to introduced plants that reproduce prolifically, interfering with the ecosystem by not allowing other plants to grow.

labiate Lip-like or having lips.

latex Milky sap.

male flower Flower bearing only male parts (stamens) and having no pistils (female parts).

naturalized Non-native species that has established in a habitat.

nectar Sugary, sticky fluid produced by various organs of a plant, but generally the flowers, to attract pollinators such as ants and other insects.

ornamental Plant cultivated for its appearance.

perennial plant Plant living and reproducing for more than two years.

petal One of the usually brightly coloured parts of a flower surrounding the reproductive organs.

pistil Female ovule-bearing part of a flower, including the stigma, style and ovary.

pollen Fine, powder-like material, the grains of which contain male reproductive cells of seed plants.

raceme Inflorescence having stalked flowers arranged singly along an elongated, unbranched axis, with flowers at the bottom opening first.

seed Sexual reproductive structure in which embryo is housed.

seedling Very young plant growing from recently germinated seed.

sepal Individual part of protective layer around a flower, which is either leaf-like or petal-like.

stamen Male reproductive structure that has a filament (stalk) and pollen-bearing anther.

stigma Top part of pistil that receives pollen.

style Part of pistil between stigma and ovary.

whorl One of several layers around central axis in a flower.

Fire Cracker Flower ■ *Crossandra infundibuliformis*

DESCRIPTION Evergreen shrub that grows to 1m tall. It has a compact form with elliptic, glossy green leaves. Attractive yellow to orange flowers arranged in dense clusters on top of a spike in spring–summer, and sporadically throughout the year. Flowers are day blooming and attract insect pollinators. **HABITAT** Native to southern parts of India, Sri Lanka, Bangladesh and Central Africa. Occurs in forest thickets and foothills in its

natural habitat, and commonly planted in gardens across India. Suited to a position in full sun, but also does well in partial shade. **USES** Dwarf cultivars used popularly as a groundcover in gardens. Mostly cultivated for its showy ornamental flowers. In South India it is a popular flower offered in temples and known by its local name, *Kanakambaram*. **ETYMOLOGY** Name is derived from the Greek word *krossoi*, fringe, and *aner*, male or anthers, referring to its fringed stamen. The epithet *infundibuliformis* describes the shape of the flower – like a funnel or trumpet.

Blue Sage ■ *Eranthemum pulchellum*

DESCRIPTION Evergreen spreading shrub with dense foliage. Can grow to 1–1.5m tall. Leaves elliptic, large, dark green and veined. Popular garden plant for its attractive foliage

and flowers. Trumpet-shaped, blue-violet flowers appear abundantly in spring, February–March. They are arranged densely on a spike, clustered near the leaves. **HABITAT** Native to India and western China. Occurs naturally in the Himalayan foothills, and prefers a spot in partial shade. **USES** Cultivated in gardens for its foliage as well as its ornamental flowers. Good filler plant under large trees and often used as a groundcover. Host plant for butterflies and suitable for planting in butterfly gardens. **ETYMOLOGY** The epithet *pulchellum* is derived from the Greek *pulcher*, pretty or beautiful, which aptly describes its blooms.

Shrimp Plant ■ *Justicia brandegeeana*

DESCRIPTION Small shrub with evergreen foliage and attractive, unusually shaped flowers. Grows upright to a height of 1m. Flowers borne on a colourful spike that has a curved, shrimp-like form, hence the plant's common name. Flowers small and inconspicuous, white in colour and appear in winter. Lower part of white petals striped with mauve to serve as a guide to its nectaries. Showy yellow-orange bracts covering small, tubular flowers are designed to attract pollinators. In its natural habitat, known to attract hummingbirds. **HABITAT** Native to Mexico, and naturalized and cultivated widely in the Indian subcontinent as a houseplant. Can survive in full sun and thrives in partial shade. **USES** Not common in public parks and gardens. Popular houseplant and does well in pots and containers. In its native region, used to treat stomach ailments and wounds. **ETYMOLOGY** The genus *Justicia* is named after the Scottish botanist James Justice.

Yellow Shrimp Plant ■ *Pachystachys lutea*

DESCRIPTION Small evergreen perennial shrub that grows to about 1m tall. Dark green leaves ovate, broad and tapering, providing dense, attractive foliage. Flowers similar to those of the Shrimp Plant (see above), but have an upright form. Spike of bright yellow bracts protects narrow white flowers that peep out from behind bract. Bright, conspicuous bracts serve to attract the attention of pollinators. Flowers appear in warm summer months. **HABITAT** Native to Costa Rica, Venezuela to Peru, and commonly cultivated throughout India as a houseplant. Suitable for growing in full sun, and also does well in partial shade. **USES** Mainly cultivated as an ornamental houseplant. **ETYMOLOGY** Name of the genus, *Pachystachys*, is derived from the Greek *pachys*, thick, and *stachys*, spike, referring to the shape of the inflorescence. The epithet *lutea* refers to its yellow colour.

Nongmankha ■ *Phlogacanthus thyrsiformis*

DESCRIPTION Tall shrub that grows upright to 2.5m in height. It has attractive foliage of large, tapering leaves. Its numerous species are found in the north-east states of India. A common variety grown many home gardens of Manipur is locally known

as *Nongmankha* and can be distinguished by its yellow-orange flowers. *P. parviflorus* is a similar species with red flowers. Tubular shaped flowers are borne on densely packed spikes and appear in winter. **HABITAT** Native to the subtropical Himalayas from Garhwal to Bhutan, northeastern India, Myanmar, Indo-China and south China. Cultivated popularly in kitchen gardens in Manipur. **USES** Leaves used in home remedies to treat colds and coughs. Flowers used to treat skin ailments. All parts of the plant are known to have medicinal properties. It is an analgesic, anti-inflammatory and antioxidant. **ETYMOLOGY** The epithet is derived from the arrangement of the flowers, which is a thyrsus-like cluster, with a central spike and flowers on side branches.

Mexican Blue Bell ■ *Ruellia tweediana*

DESCRIPTION Evergreen hardy perennial shrub with a sprawling, spreading habit. Long, narrow leaves sharply tapered and sparse. Purple-blue flowers appear sporadically through warm summer months. They are trumpet shaped and arranged loosely on panicles. A dwarf version known as var. *compacta* has blue-and-pink flowers. Flowers are short-lived and wither in a day. They are visited by butterflies and bees. **HABITAT** Native to Mexico. Hardy plant that adapts to most conditions. Widely cultivated in India as a garden plant. Does well in full sun as well as partial shade. **USES** Sprawling habit makes it suitable for cultivation as a groundcover in gardens. Vigorous plant that can take over a garden and quickly become invasive.

King's Mantle ■ *Thunbergia erecta*

DESCRIPTION Evergreen shrub with spreading, scandent branches. Can grow to 2.5m in height. Dark green, glossy leaves 2–5cm long, dense and make attractive foliage. Prominent trumpet-shaped flowers purple-blue in colour with pale yellow throat. They are borne solitarily and sometimes in pairs. Blooms profusely in spring and sporadically

during monsoon season. **HABITAT** Native to tropical parts of western Africa. Introduced in India and cultivated widely as a garden plant. Thrives in sandy, loamy soil in a partially shaded location or full sun. **USES** Grown for its glossy foliage as well as its ornamental flowers. Can be trimmed and maintained as a hedge or edge planting in planter beds. **ETYMOLOGY** Genus named after Swedish naturalist Carl Peter Thunberg, and the epithet *erecta* describes the upright growth habit of its form.

Bengal Clock Vine ■ *Thunbergia grandiflora*

DESCRIPTION Quick-growing evergreen climber that becomes woody over time. Can grow to 15m in height. Bright green leaves with irregular lobes are dense and vary in size and shape. Pale blue flowers with petals that are nearly white on the outside borne in loose, drooping clusters. Flowers profusely in summer and sporadically throughout the year. Pollinated by insects and commonly visited by bees. **HABITAT** Native to India in the eastern Himalayas from Nepal to Myanmar and parts of China. Natural habitat is along water courses and forests margins. Thrives in well-drained, sandy soil with plenty of sunlight. **USES** Cultivated as a garden ornamental for its showy blue flowers. Suited to training over fences and screens. **ETYMOLOGY** Genus named after Swedish naturalist Carl Peter Thunberg. The epithet *grandiflora* refers to the large, prominent flowers.

Scarlet Clock Vine ▪ *Thunbergia coccinea*

DESCRIPTION Vigorous climber with dense evergreen foliage of deeply veined leaves. A prominent feature is its scarlet flowers, borne in long, drooping racemes. Two-lipped petals orange-red in colour with yellow throats. They appear in autumn and flower until spring, September–March. **HABITAT** Native to lower ranges of the Himalayan region from Nepal to Myanmar. In its natural habitat, found along mountain slopes. Thrives in tropical and subtropical climates and moist soil conditions. **USES** Cultivated as an ornamental in gardens. Best trained on to a pergola structure, which allows the drooping flowers to form a sky of scarlet blooms. **ETYMOLOGY** Genus named after Swedish naturalist Carl Peter Thunberg. The epithet *coccinea* refers to the scarlet-coloured flowers.

Lipstick Tree ■ *Bixa orellana*

DESCRIPTION Small evergreen tree or shrub with a spreading crown, growing to 6m in height. Dense dark green foliage with large, heart-shaped leaves. White-pink flowers appear in spring and bloom throughout summer. They are borne in clusters and

are pollinated by bees. Numerous seeds are produced inside a pod. **HABITAT** Native to Central and South America. Natural habitat is in tropical forests and along sea coasts. Widely cultivated in India and thrives in warm, humid climate. **USES** Ornamental shrub planted in gardens for its glossy foliage and clusters of red fruits. Its leaves, roots and seeds are known to be astringent and purgative, and it is used in herbal remedies. Annatto is a popular red-coloured food dye obtained from the fruits. Seed cover is used in cosmetics and food colouring. Many parts of the tree are used in India for making traditional medicines. **ETYMOLOGY** Genus name is derived from the vernacular term for the plant in South America. The epithet is named in honour of the Spanish explorer Francisco de Orellano.

Tuberose ■ *Polianthes tuberosa*

DESCRIPTION Perennial with tuberous roots, and long, narrow leaves that bend gracefully. Known for its fragrant white flowers, which are funnel shaped and closely packed on a tall spike. Most fragrant in the evenings, when the buds open, and known to attract insect pollinators like nocturnal moths. Flowers appear in summer months. **HABITAT** Native to Mexico, and introduced to India via Europe. Popularly cultivated in Indian gardens, and also for floriculture. Thrives in hot, moist climates. **USES** Fragrant extracts of floral oils used in perfume making. Popular as a cut flower for floral arrangements, and also used in strings for ceremonial garlands. Recommended as a plant for moonlight gardens, since its beauty and fragrance can be best appreciated at night. **ETYMOLOGY** Genus name is derived from the Greek *polios*, white, and *anthos*, flower, referring to its white blossoms. The epithet *tuberosa* describes its tuberous roots.

New Zealand Flax ■ *Phormium tenax*

DESCRIPTION Evergreen perennial shrub with attractive grey-green foliage. Large shrub that can grow to 2–2.5m in height. Leaves long and tapered with sharp-pointed ends. Flowers borne on long stems arranged in panicles. Dull red or orange flowers tubular in shape and have two phases. In the male phase the colours are bright orange, and after pollen has been dispersed they fade to pale yellow, during the female phase. They appear in summer and are known to be visited by birds and honey bees. **HABITAT** Native to New Zealand and Norfolk Island. Thrives in subtropical climate with moist soil conditions. Not common in India. Cultivated in higher altitudes of Shimla and Kathmandu in Nepal. **USES** Ornamental shrub with attractive form and foliage. In its native region, fibre is used in weaving fabrics and baskets. Flowers are rich in tannin. **ETYMOLOGY** Genus name is derived from *phormion*, mat, and the epithet *tenax* means tough.

Day Lily ▪ *Hemerocallis fulva*

DESCRIPTION Perennial shrub with clump-forming habit. Grows from tuberous rhizomes to 0.5m tall. Slender, graceful leaves grass-like and shed in winter. Orange-yellow flowers trumpet shaped with a wide mouth, and borne solitarily or in pairs on a tall stalk. They bloom in summer, in June–November, and last for a day. **HABITAT** Native to East India, China, Japan and Taiwan. Found along streams and on forests grounds. Thrives in full sun and partially shaded locations. **USES** Grown as a groundcover for its dense foliage and ornamental flowers. In east Asia, flowers are eaten raw or dried. Leaves and flowers used in traditional medicines. **ETYMOLOGY** Genus name is derived from the Greek *hemera*, day, and *kallos*, beauty, in description of the beautiful flowers. The epithet *fulva* refers to the orange-coloured blossoms.

Hyacinth ▪ *Hyacinthus orientalis*

DESCRIPTION Annual herbaceous shrub growing from a bulb to 0.3m tall. Strap-like leaves are upright and appear at the same time as the flowers. Flowers highly fragrant, packed densely on a spike and pollinated by bees. Colours range from pink and mauve, to purple, and bloom in spring, in March–April. Bulbs become dormant in summer after the

leaves drop. **HABITAT** Native to southern Turkey and Israel. Not commonly cultivated in India, and grown as a winter seasonal. **USES** Garden ornamental prized for its fragrant blossoms that last long as cut flowers. Essential oil from flowers used in perfumes. **ETYMOLOGY** Genus is named after the Greek legend of *Hyakinthos*, which refers to a rebirth of vegetation, and the epithet refers to its origins in the Orient.

Johnson's Amaryllis ■ *Hippeastrum reginae*

DESCRIPTION Evergreen or seasonally dormant herb growing from a bulb. Dark green, strap-like leaves provide lustrous foliage throughout the year. Large red flowers have white star-shaped marking in the throat. They are trumpet shaped, with a tubular base

and spreading lobes, and borne on a hollow stalk growing directly from the bulb. They bloom in spring, April to early May. Many cultivars are grown from crimson-flowering *H. reginae*, with flower colours in shades of white, pink, red or orange. **HABITAT** Native to northern parts of South America. Cultivated widely in most parts of India. Does well in full sun, as well as in partially shaded locations. **USES** Cultivated as an ornamental garden plant. Flower stalks often cut and arranged in vases. **ETYMOLOGY** Genus name is derived from the Greek *hippeaus*, knight, and *astron*, star. Often incorrectly referred to as 'amaryllis'.

African Lily ■ *Agapanthus africanus*

DESCRIPTION Perennial lily growing from rhizomes, with long, dark green leaves that are strap shaped. Can grow to 1m in height. Known for its lavender-blue flowers, which bloom in summer, April–September. Some varieties also have white flowers. Numerous trumpet-shaped flowers arranged in a dense cluster like a round globe, borne on a tall stalk. **HABITAT** Native to southern parts of Africa in the Cape Province, and cultivated in cooler climes of India. Prefers a sunny position and thrives in moist soil conditions like edges of waterbodies. **USES** Ornamental garden plant cultivated for its beautiful flowers. Also used in cut-flower arrangements. **ETYMOLOGY** Genus name is derived from *agape*, love, and *anthos*, flower.

Amazon Lily ■ *Eucharis amazonica*

DESCRIPTION Evergreen lily growing from a bulb. Attractive foliage glossy green, lance shaped or elliptic, and deeply veined. Clusters of white flowers appear in spring, in February–March. They are similar to the flowers of *Narcissus*, star shaped with a prominent central corona. However, flowers of the Amazon Lily are drooping in form and arranged at the end of a long stalk. They are intensely fragrant during the day. **HABITAT** Native to Peru, and cultivated as a garden ornamental in India and Sri Lanka. In its natural habitat, found growing under the cover of rainforests. Thrives in shady locations with moist soil conditions. **USES** Grown for its fragrant white flowers, doing well as a pot plant. **ETYMOLOGY** Genus name *Eucharis* in Greek means grace or charm, referring to its flowers. The epithet *amazonica* refers to its origin in the Amazon.

Bush Lily ■ *Clivia miniata*

DESCRIPTION Evergreen perennial growing from rhizomes. Compact form with prominent flower clusters. Dark green leaves long and narrow, and provide attractive foliage. Leaves generally about 7.5cm wide and 45cm long. Orange-red flowers funnel shaped with yellow centres. Numerous flowers arranged in a tight cluster on top of a long stalk. Flowers appear in spring. **HABITAT** Native to South Africa and Swaziland, and widely cultivated in cooler regions of India, Nepal and Pakistan. Native habitat ranges from coastal to subtropical forests. **USES** Grown as an ornamental garden plant in India. In its native region, the toxic rhizomes are used for medicinal applications. All parts of the plant are poisonous. **ETYMOLOGY** Genus named after Lady Charlotte Florentina Clive by a Kew botanist. The epithet name *miniata* refers to the vermilion-red colour of the flowers.

Daffodil ■ *Narcissus tazetta*

DESCRIPTION Bulbous perennial with long, narrow leaves growing to 0.5m tall. Flowers are a combination of white, cream or yellow, and star shaped with a prominent central corona. Typically arranged in radiating cluster from a long stalk, they are intensely fragrant. Flowers appear in spring. *Narcissus* are classified into divisions based on the shapes of the flowers. Some important species are *N. jonquilla*, *N. tazetta*, *N. cyclamineus* and *N. poeticus*.

HABITAT Native to Mediterranean region from Portugal to Turkey. Naturalized across the Indian subcontinent as well as the Middle East, Central Asia, Iran and many other parts of the world. Introduced in the subcontinent and cultivated as a winter seasonal in most parts of the region. Grows well in all kinds of soils – sandy, loamy and clay. Prefers a sunny location. **USES** Cultivated as an ornamental winter seasonal in Indian gardens. Flowers popular in cut-flower markets. Essential oils from species like *N. jonquilla* are used in perfumery.

Rain Lily ■ *Zephyranthes candida*

DESCRIPTION Evergreen perennial rising from a bulb, and growing to 0.3m tall. Small, grass-like herb with narrow, dark green leaves. White flowers funnel shaped and borne solitarily on a slender stalk. They appear at the onset of the monsoon and bloom through

the rainy season in June–September, hence the name Rain Lily. Other species commonly seen in subcontinent gardens are Z. *rosea* with small, dark pink flowers, Z. *citrina* with small yellow flowers, and Z. *grandiflora* with large, light pink flowers. **HABITAT** Native to La Plata, Argentina, but widely naturalized across the globe. Thrives in moist soil conditions and planted in full-sun location. Widely cultivated in gardens throughout the Indian subcontinent. **USES** Planted as a groundcover in beds for its ornamental flowers. **ETYMOLOGY** Genus name is derived from the Greek *zephyros*, west wind, and *anthos*, flowers. The epithet *candida* refers to its pure white flowers.

Spider Lily ■ *Crinum asiaticum*

DESCRIPTION Evergreen bulbous perennial with long, sword-shaped leaves tapering to the ends. Can grow to 2m in height. Bright green foliage grows upright. Flowers similar to those of *Hymenocallis*, with white flowers clustered at the end of a long stalk. Petals thin and spider-like, surrounding prominent stamens on long filaments. Flowers fragrant at night and pollinated by nocturnal moths. Flowers profusely in summer and sporadically almost year round. HABITAT Native to tropical parts of Asia, and widely cultivated in Indian parks and gardens. Adapts to most conditions and thrives in moist areas near waterbodies. Suitable for growing in a wide range of soils, from sandy and medium, to heavy. Prefers a sunny location. USES Cultivated as an ornamental garden plant. ETYMOLOGY Genus name is derived from the Greek *krinon*, lily, and the epithet *asiaticum* refers to its origins in Asia.

Swamp Lily ■ *Crinum moorei* 'Album'

DESCRIPTION Evergreen bulbous perennial with long, broad leaves tapering at ends. Bright green foliage grows upright to 1–1.5m in height. Flowers showy, large and white, opening at night with a heady fragrance. Flowers appear in the monsoon season. HABITAT Native to South Africa, and not common in Indian gardens. Prefers a location in partial shade and does well in moist conditions. Grows in all soils – sandy, loamy and so on. USES Cultivated as a garden ornamental plant. ETYMOLOGY Genus name is derived from the Greek *krinon*, lily, and the epithet is named after Dr D. Moore, who was a director at the Botanical Garden in Dublin.

Water Canna ■ *Thalia dealbata*

DESCRIPTION Perennial shrub often used as a bog or marginal aquatic plant, growing upright to a height of 1–1.5m. Leaves similar to those of Canna (see p. 50). They are lance shaped and bright green. Flowers of different shades of purple appear in summer and are arranged loosely on long stalks. Both the foliage and the flowers are ornamental. **HABITAT** Native to southeastern USA and Mexico. Grows near margins of waterbodies and swamps. Prefers a position in full sun. **USES** Attractive plant for growing in ponds, and can also be used in constructed wetlands and reedbeds for treating waste water. **ETYMOLOGY** Genus is named in honour of German physicist and botanist Johannes Thal. The epithet *dealbata* means covered in white powder, referring to the foliage, which is coated with it.

Calla Lily ▪ *Zanthedeschia aethiopica*

DESCRIPTION Small shrub with glossy green foliage that can be deciduous or evergreen. Grows upright to 1m tall. Leaves broad, and lance or arrow shaped. Prominent large white flowers appear in spring, in March–April, and are borne solitarily on a long stalk. Similar to all flowers of the arum family, they comprise a bract, or spathe, and a spike, or spadix. Creamy white bract unfurls around spike, creating an unusual attractive shape. Flowers faintly fragrant and pollinated by insects like bees.

HABITAT Native to South Africa and Lesotho, and cultivated as an exotic garden plant in the Indian subcontinent. Thrives in cooler climate conditions and higher altitudes, such as those in Kathmandu in Nepal. Versatile plant also known to do well in coastal areas. **USES** Grown as an ornamental garden plant for both its showy foliage and flowers. Flowers are long lasting and can be used in cut-flower arrangements. The species adapts to wet marshy conditions and can be grown as a marginal water plant. **ETYMOLOGY** Genus is named after an Italian botanist, Giovanni Zantedeschi. The epithet *aethiopica* refers to the plant's region of origin in Ethiopia or Africa in general.

Peace Lily ▪ *Spathiphyllum wallisii*

DESCRIPTION Evergreen perennial shrub with dark green, glossy foliage. Grows from rhizomes to 0.6m tall. Leaves oblong and sharply tapered on long stems. Flowers typical of the arum family, with a large white bract framing a cream-white, fleshy spike on which the flowers grow. They appear in summer.

HABITAT Native to Colombia and Venezuela. Commonly cultivated in most parts of the Indian subcontinent. Suited to semi-shade areas with moist, well-drained soil. **USES** Grown as an ornamental houseplant, and suitable for planting as a groundcover in gardens. Also a popular indoor plant, ideally placed near a brightly lit window with natural light and ventilation. Recently recognized as an air-cleaning plant. **ETYMOLOGY** Genus name *Spathiphyllum* means leafy spathe, and the epithet is named after plant collector Gustav Wallis.

Flamingo Flower ▪ *Anthurium andraenum*

DESCRIPTION Evergreen perennial shrub with attractive leathery, glossy green leaves. Small in size, growing to 0.6m. Leaves heart shaped with sharply pointed tips, on long stems. Flowers showy with a prominent, brightly coloured spathe or sheath that encloses the flower, and a fleshy spike or spadix on which the minute flowers are arranged. Colours of the waxy spathe can be white, pink, purple and even pale yellow. Flowers

appear sporadically throughout the year. **HABITAT** Native to southwestern Colombia. Common houseplant in cooler regions of India like the north-east states. Available in plant nurseries as exotics in extreme climate like that of Delhi. Requires good drainage, provided by coconut husk and mulch, to avoid root rot. **USES** Cultivated as an exotic houseplant. Mostly grown in containers and can be kept as an indoor plant. Flowers are long lasting and for that reason this is a popular cut-flower plant. **ETYMOLOGY** Genus name *anthurium* means tail flower, and the epithet refers to French botanist Edouard F. Andre.

Impatiens ■ *Impatiens walleriana*

DESCRIPTION Perennial herbaceous plant of low, spreading habit, which grows to 0.3m tall. Leaves simple and arranged alternately. They are oblong with tapered ends and serrated edges. Flowers borne solitarily, blooming profusely in multiple colours – red, pink, orange and white. They are saucer shaped, comprising five petals with a curved tube at the back that bears nectar. **HABITAT**
Native to East Africa, from Tanzania to Mozambique, and widely cultivated in the Indian subcontinent. In hot dry regions such as Delhi, grown as a winter seasonal. Suited to growing in partial shade or full sun. **USES**
Ornamental shrub that can be grown as a groundcover. Usually grown as a winter seasonal flowering plant. **ETYMOLOGY**
Genus name means impatient, referring to the discharge of seeds from ripe pods that burst open under pressure. The epithet *walleriana* refers to a botanist named Waller, who was a collector of plants in Britain.

Banana ■ *Musa* x *paradisiaca*

DESCRIPTION Large herb that grows tall and upright, with multiple fronds spreading out from the top like in a palm. Stem is made up of overlapping leaf sheaths with large, oval-shaped leaves that can be more than 1m long. Solitary, drooping flowers borne in summer, enclosed in dark purple-maroon bract. Comprises multiple layers of bracts, with each
layer containing 2–3 rows of flowers that are tubular in shape. Cultivated varieties of banana do not require pollination, but the flowers of wild species release a strong odour to attract nocturnal bats for pollination. There are more than 60 known species of *Musa* and many of them grow in the Indian subcontinent. Many wild species also exist, including the recently discovered *Musa arunachalensis*. **HABITAT** Native to tropical parts of Asia and Africa, and cultivated widely in India as a food crop. Thrives in moist, humus-rich soil in full sun. **USES** Cultivated for its nutritious fruits, and stem and flowers are also edible. In East and South India, stems and flowers are cooked and eaten. Good source of iron. Leaves used as platters to serve food in many parts of the Indian subcontinent. **ETYMOLOGY** Genus name is assumed to be in honour of the ancient Greek botanist and physician Antonia Musa.

Asian Barberry ■ *Berberis asiatica*

DESCRIPTION Spiny evergreen shrub with multi-branching habit. Grows to 3m in height. Leaves serrated with sharp-toothed edges. Small yellow flowers densely packed at end of a spike. Flowers profusely in spring, in February–May. **HABITAT** Native to the Himalayan range. Found in India, Nepal, Bhutan and parts of Tibet. In its natural habitat, *Berberis* grows on shaded, north-facing slopes. It is cultivated in gardens in cooler climates, such as those of Bhutan, Sikkim, Kashmir and Kathmandu. **USES** Grown as an ornamental plant in gardens and parks, and also suited to being trained as a hedge due to its spiny characteristics. Edible fruit is dried and eaten like a raisin. Most parts of the plant are used medicinally and have alkaline properties.

Nepal Barberry ■ *Mahonia nepaulensis*

DESCRIPTION Evergreen shrub or small tree that grows to 1–4m high. Dark green leaves are glossy with sharply serrated edges. Prominent spikes of yellow flowers are fragrant

and appear in spring, in March–April. Small flowers are numerous and arranged densely on a spike. **HABITAT** Native to high altitudes (1,500–3,400m) of the Himalayan range of India, Nepal and Bhutan, in oak and rhododendron forests. Also found in the Western Ghats. Thrives in partially shaded areas but also does well in full sun. **USES** Grown in gardens for its ornamental flowers. Fruits edible and can be eaten raw or cooked. Berberine compounds contained in the plant are known to have antibacterial and anti-inflammatory properties. **ETYMOLOGY** Named after American Irish botanist Bernard M'Mahon, who introduced the plant to America. The epithet *napaulensis* refers to its origins in Nepal.

Madhavi ▪ *Hiptage benghalensis*

DESCRIPTION Large woody climber that grows to 10m in height. Large, oblong leaves are evergreen and have tapered tips. Flowers appear in spring, in February–May, and are arranged in loose clusters. Petals white with yellow centres, and frilly at edges. They are intensely fragrant and attracts bumblebees and other insects. Fruits are peculiar, attached with three wing, and when dry are carried by the wind. **HABITAT** Native to India and Southeast Asia. Found in the Eastern and Western Ghats and parts of Northeast India. Widely cultivated in Indian gardens. **USES** Grown in gardens for its fragrant flowers. Leaves and aromatic bark have medicinal properties and are used in herbal remedies. **ETYMOLOGY** Genus name is derived from the Greek *hiptamai*, to fly, referring to its winged fruits. The epithet name *benghalensis* means from Bengal.

Spray of Gold ▪ *Galphimia gracilis*

DESCRIPTION Medium-sized evergreen shrub that grows to 1.5m tall. Its rounded form is covered in fine foliage of small, narrow leaves. Star-shaped flowers are borne in clusters. Profuse in summer, they cover the entire plant in yellow blossoms. **HABITAT** Native to tropical parts of America, Mexico and Guatemala. Cultivated in India and other countries of the subcontinent as a popular garden plant. Thrives in full sun. **USES** An ornamental garden plant, used as a filler in plant beds. Also suited to planting in containers and pots. **ETYMOLOGY** Genus name *Galphimia* is an anagram of *Malpighia*, and the epithet *gracilis* means slender or graceful.

Wax Begonia ■ *Begonia x semperflorens cultorum*

DESCRIPTION Annual shrub with bushy, dense foliage and fleshy stems, which grow to 0.3m tall. Waxy leaves broad and heart shaped, and arranged alternately. They are glossy green or bronze-red in colour. Flowers are pretty, in shades of white, pink and red, with yellow stamens in the centre. They have four petals, which are arranged in a cruciform shape. They bloom in cold months in the plains, December–March, and summer months

in the hills, and there are single- or double-flowering varieties, borne in clusters. **HABITAT** Hybrid of the *semperflorens* group, thriving in cool subtropical climates. Cultivated widely in India as a winter seasonal. **USES** Ornamental seasonal flowering shrub used as a bedding plant or in containers. Prefers a position in partial shade. **ETYMOLOGY** Genus is named after Michel Begon, a governor of French Canada and a collector of plants. Group name *semperflorens* means ever flowering.

African Tulip Tree ■ *Spathodea campanulata*

DESCRIPTION Evergreen tree of medium size and with rounded crown. Grows to 20m in height. Glossy green leaves oval shaped with tapered ends. Large, scarlet-orange flowers appear in spring, in February–March, and again in autumn, in October–November. They are arranged in a dense, cone-like cluster, and are bell shaped or campanulate, like a tulip, hence the common name African Tulip Tree. They attract nectar-feeding birds and bees as pollinators. **HABITAT** Native to tropical parts of Africa. Widely cultivated in the Indian subcontinent and does best in hot, humid regions. **USES** Cultivated as an ornamental tree in parks and gardens. Also seen commonly as an avenue tree in South India, where it grows to its optimum height. In its native region, bark and leaves are used to treat wounds, as it is known to possess antibacterial properties. **ETYMOLOGY** Genus name is derived from the Greek *spathe*, referring to the broad blade-like calyx. *Campanulata* describes the bell-shaped flowers.

Cape Honeysuckle ■ *Tecoma capensis*

DESCRIPTION Evergreen shrub with scrambling habit. Can grow to 3m in height. Dark green foliage is dense and leaves are compound, with leaflets arranged pinnately in opposite pairs. Flowers orange-red and sometimes yellow in varieties such as *T. c.* 'Aurea' . Tubular shaped with a slight curve, they are arranged in small clusters on a spike. Flowers appear year round but are profuse in September–November, and are visited by insects such as ants, bees and butterflies, and sunbirds for nectar. **HABITAT** Native to South Africa. Commonly cultivated in gardens throughout India. Thrives in full sun and also does well in partially shaded areas. **USES** Cultivated for its attractive foliage and flowers, which attract birds and insects. Suited to planting in butterfly gardens. Can be trimmed and maintained as a hedge or border plant. In its native region in Africa, bark is known to be used in traditional herbal remedies. **ETYMOLOGY** Genus name is derived from the Mexican term *tecomaxocitl*, which was applied by indigenous Mexican people to plants with tubular flowers. The epithet *capensis* means from the Cape.

Chinese Trumpet Vine ■ *Campsis grandiflora*

DESCRIPTION Deciduous climber with vigorous growth that can reach to 10m in height. Climbs and clings by means of aerial roots. Glossy green leaves compound with serrated edges. Flowers bloom in early summer, in April–October. They are borne in loose, drooping

clusters and attract sunbirds and beetles, which eat the petals. **HABITAT** Native to China. Cultivated widely in India and thrives in cooler climes of the foothills. Thrives in full sun and tolerates partially shaded locations. **USES** Ornamental climber grown in gardens and suitable for training over pergolas and trellises. **ETYMOLOGY** Genus name is derived from the Greek *kampe*, bent, referring to the curved shape of the flower. The epithet name refers to the large-sized flowers.

Garlic Vine ■ *Mansoa alliacea*

DESCRIPTION Vigorous climber that grows to 3m tall. Attractive glossy green leaves are aromatic and smell of garlic when crushed. Lilac flowers with dark purple tips are trumpet shaped and bloom profusely, covering the entire vine when in bloom. Flowers bloom twice a year, in spring and autumn, and attract bees and butterflies. **HABITAT**

Native to the Amazon forests of Brazil and Peru in South America. Thrives in warm, moist climate conditions and widely grown in India. **USES** Cultivated as an ornamental climber, and trained over fences and arbours. Leaves, bark and roots used medicinally and known to be analgesic and anti-inflammatory. In its native region, used to treat coughs, colds and fevers, and is even attributed with magical properties.

Golden Shower Vine ■ *Pyrostegia venusta*

DESCRIPTION Evergreen, quick-growing climber grown for its profuse ornamental flowers. Can grow to 12m tall. Flame-orange flowers bloom in spring, in February–April, borne in long, drooping clusters, and are tubular in shape. They are pollinated by sunbirds and also attract bumblebees. **HABITAT** Native to southern Brazil, north Argentina, Bolivia and Paraguay. Cultivated widely in gardens throughout India. Thrives in a sunny location. **USES** Cultivated as an ornamental garden plant, and used for training over fences and trellises. In its native region, used medicinally to treat infections and respiratory diseases. **ETYMOLOGY** Genus name is derived from *pyro* for flame and *stege* for covering due to the orange flowers that cover the entire plant. The epithet *venusta* means pleasing.

Jacaranda ■ *Jacaranda mimosifolia*

DESCRIPTION Deciduous, medium-sized tree with a light, spreading crown that sheds before the flowers bloom. Small leaves compound and closely resemble those of *Delonix regia*. Mauve-purple flowers bloom profusely in March–May and cover the entire crown of the tree. Trumpet-shaped flowers arranged in loose clusters and pollinated by bees and other insects. **HABITAT** Native to South America – Argentina and Paraguay. Introduced in India from Brazil and widely cultivated throughout the subcontinent. Suited for planting in the plains, and also does well in higher altitudes of the foothills. **USES** Cultivated as an ornamental tree in Indian parks and gardens. Also suited for planting as an avenue tree. **ETYMOLOGY** In Paraguay, *jacaranda* means fragrance – the timber of the tree is fragrant when cut. The epithet *mimosifolia* refers to the fine, mimosa-like leaves.

Nile Trumpet ■ *Markhamia lutea*

DESCRIPTION Evergreen tree with narrow crown. Grows upright to 10m in height. Large, oval leaves have pointed ends and a rough, coarse texture. Yellow flowers trumpet shaped and bloom sporadically throughout the year. **HABITAT** Native to East Africa, in

the region of Ethiopia, Kenya, Tanzania and Uganda. Cultivated widely in India. Thrives in moist climate with heavy rainfall, but also adapts to hot, dry regions. **USES** Grown in Indian parks and gardens as an ornamental tree for its yellow flowers that bloom year round. Quick-growing tree, suited for planting as screens and windbreaks. **ETYMOLOGY** Genus named after the English geographer Sir Clement Markham, and the epithet *lutea* refers to the golden-yellow flowers.

Sausage Tree ■ *Kigelia pinnata*

DESCRIPTION Medium-sized tree with spreading crown of dense foliage. Grows to 15m in height. Leaves oblong in shape with a coarse texture. Large, maroon-red flowers appear in March–May, drooping down from the tree at the end of a long, rope-like string. Flowers bloom at night and emit an odour that attracts bats, and start to fall in the morning.

Oblong fruits shaped like a sausage, from which the common name Sausage Tree originates, develop after the blossoms. **HABITAT** Native to South Africa, and widely cultivated in India. **USES** Due to its dense crown, suited to planting as a shade tree. Often planted along urban roads as an avenue tree. **ETYMOLOGY** Genus name is derived from its Mozambique name *Keigeli-Keia*. The epithet refers to its pinnate leaves.

Yellow Bells ▪ *Tecoma stans*

DESCRIPTION Large shrub or small tree with dense evergreen foliage. Grows to 3–4m tall. Bright green leaves pinnately compound with serrated edges. Yellow flowers funnel shaped and scented, and arranged in prominent clusters. They attract bees, butterflies and sunbirds. **HABITAT** Native to Central and South America. Hardy plant widely cultivated in India, especially along roadsides and medians. **USES** Cultivated in gardens and parks for its showy flowers. Also commonly used as a hedge plant. Utilized in cultivating biodiversity parks since it attracts many insects and birds. Used in medicines to treat stomach disorders and diabetes. **ETYMOLOGY** Genus name is derived from the Mexican term *tecomaxochitl*, which was applied by indigenous Mexican people to plants with tubular flowers. The epithet *stans* refers to its upright growth.

Yellow Trumpet Tree ▪ *Tabebuia aurea*

DESCRIPTION Medium-sized tree with a semi-deciduous crown with a rounded form, and often a twisted trunk. Grows to 8m in height. Grey-green leaves elliptical in shape with blunt ends, and arranged palmately in groups of five leaflets. Flowers deep yellow and shaped like trumpets, and appear in clusters. They bloom in spring, February–April, and are visited by bees. Nectar of *Tabebuia* flowers is known to be an important food source for bees and hummingbirds. **HABITAT** Native to South America from Peru to Paraguay, and widely cultivated in India. **USES** Ornamental tree planted in parks and gardens for its attractive flowers. Root system is shallow and the tree can topple easily in a storm, which makes the species unsuitable for planting as an avenue tree. **ETYMOLOGY** Genus name is derived from vernacular Brazilian name, and the epithet *aurea* refers to the golden-yellow flowers.

Elephant Creeper ■ *Argyreia nervosa*

DESCRIPTION Large climber with attractive evergreen foliage. Grows to 9m in height. Broad, heart-shaped leaves simple and arranged alternately. They are silver-grey on the underside and covered in fine hairs. Trumpet-shaped flowers mauve with dark purple throats and resemble those of *Ipomoea*. Flowers bloom in August–November, and attract birds and bees. **HABITAT** Native to India, and found in Assam, Bengal, Bihar and southern India. Cultivated mainly as a medicinal plant and often in gardens

as a specimen plant. **USES** Quick-growing plant grown as an ornamental for its attractive foliage and flowers. Important medicinal plant – all parts used in traditional medicine systems such as Ayurveda, and considered to be rejuvenating, analgesic, anti-inflammatory and antioxidant. **ETYMOLOGY** Genus name is derived from the Greek *argyraeus*, or silver coloured, which refers to the leaves. The epithet *nervosa* refers to the hallucinogen compound contained in the seed.

Railway Creeper ■ *Ipomea cairica*

DESCRIPTION Vigorous climber with evergreen palmate leaves and slender stems. Grows to 5m in height. Leaves deeply lobed, divided into five segments and arranged alternately. Trumpet-shaped flowers purple to white, borne solitarily or in small clusters. They bloom

intermittently throughout the year and attract insect pollinators. **HABITAT** Native to tropical parts of Africa and Asia, including the Indian subcontinent. Found growing wild along roadsides and wasteland, and also along coastlands. **USES** Often omitted from gardens due to its invasive nature. However, it is a quick-growing climber with attractive foliage and flowers that can be trained over trellises as a screen. **ETYMOLOGY** Genus name is derived from the Greek *ips* or *ipos*, woodworm, and *nomoios*, resembling, referring to the creeping, crawling habit of the plant.

Sky-blue Cluster Vine

■ *Jacquemontia pentantha*

DESCRIPTION Slender evergreen climber with vigorous growth, growing to 3–4m tall. Leaves simple and heart shaped, and arranged alternately. Light blue or purple flowers with white centres borne in profuse clusters. They bloom intermittently throughout the year and attract bees and butterflies. **HABITAT** Native to Central America, from Florida and Mexico up to South America. Widely cultivated in India and prefers a tropical climate with a position in full sun. **USES** Quick-growing climber with attractive foliage and flowers. Can be trained over fences and trellises as a screen. **ETYMOLOGY** Genus is named after French botanist Victor Jacquemont, and the epithet *pentantha* means five flowered.

Calico Flower ■ *Aristolochia elegans*

DESCRIPTION Evergreen climber with vigorous growing habit, which becomes woody. Bright green leaves are heart shaped. Large, conspicuous flowers oddly shaped, like a pipe. Buds enclosed, like a pouch that opens up to a wide mouth, showing the purple-veined interior. Most species of *Aristolochia* flowers emit an odour to attract insect pollinators. They appear in summer and bloom until autumn. **HABITAT** Native to South America. Not commonly cultivated in the Indian subcontinent's gardens. Thrives in warm, humid regions and planted in full sun. **USES** Grown for its large, ornamental flowers, and also for its use in medicinal remedies. **ETYMOLOGY** Genus name is derived from the Greek *aristos*, best, and *locheia*, birth, referring to the medicinal use of the plant during childbirth.

Scarlet Cordia ■ *Cordia sebestena*

DESCRIPTION Evergreen tree that has a compact, rounded form. Grows to 8m in height. Leaves simple, oval shaped and with tapered ends. Attractive orange-red blossoms appear intermittently in warm summer months. Flowers trumpet shaped and arranged in small clusters. They attract birds and butterflies for pollination. **HABITAT** Native to the Caribbean region, from Florida, to Mexico in Central America, and Venezuela in South

America. Cultivated in most parts of India, and most commonly seen as a roadside tree in cities in the state of Gujarat. Tolerates saline conditions and found growing along sea coasts and in sandy soil. **USES** Cultivated as a garden ornamental, and suited for planting in small gardens due to its compact form. Also planted as an avenue tree. Fruit can be eaten raw or cooked. **ETYMOLOGY** Genus is named after German botanist Valerius Cordus. The epithet *sebestena* is derived from a similar species that grows in a town called Sebesta in historical Persia.

Cannonball Tree ■ *Couroupita guaianensis*

DESCRIPTION Large deciduous tree with a straight truck. Grows to 25–30m in height. Simple, oval-shaped leaves tapered at ends and arranged in whorls at the ends of branches. Flowers appear in February–November, in long, drooping, rope-like clusters covering trunk of tree. Petals thick and waxy, of an orange-red colour on the inside; centre is a hood-like structure comprising closely packed stamens. Highly fragrant, and fragrance intensifies at night; it attracts bees in the day and bats by night. Fruits large and round like a cannonball, hence the common name. **HABITAT** Native to South America, in the Amazon rainforests and Guiana. Planted as a specimen tree in India, it is most commonly seen in southern India, where it is considered sacred. **USES** Cultivated in parks as a specimen tree due to its curious ornamental hanging flowers and fruits. Flower is used as a votive offering and is associated with the Hindu god Shiva, as its structure is likened to the hood of the Naga snake over a Shiva Lingam. Often grown in temple complexes in South India. Floral oil from tree used in cosmetics and perfume making. Extracts from tree used in medicines, and known to have antiseptic and antifungal qualities. **ETYMOLOGY** Genus name is derived from the vernacular term in Guiana, *kouroupitoumou*, and the epithet *guianensis* means of Guiana.

Persian Buttercup ■ *Ranunculus asiaticus*

DESCRIPTION Annual herbaceous shrub that grow upright from tuberous roots, to a height of 0.4m. Leaves deeply lobed and divided into three segments, resembling parsley. Cup-shaped flowers are attractive, borne solitarily or in pairs, and occur in various colours – white, pink, yellow, orange and red. They bloom in December–March in the plains and in July–September in the hills. Many garden hybrids are available, including double-

flowering forms. **HABITAT** Native range is widespread over south-east Europe, northern Africa and south-west Asia. Cultivated widely in India, in both hills and plains. **USES** Ornamental plant grown commonly as a seasonal for its showy flowers. Also popular with florists as a cut flower. **ETYMOLOGY** Genus name is derived from the Latin *rana*, frog, due to its affinity with damp or moist conditions. The epithet *asiaticus* refers to its Asian origin.

Coral Vine ■ *Antigonon leptopus*

DESCRIPTION Slender evergreen vine that grows to 6m tall and climbs by means of curling tendrils. Heart-shaped leaves simple and arranged alternately. Bright pink flowers have yellow centres, appear in drooping clusters in April, and bloom throughout the summer months. They attract bees and butterflies. **HABITAT** Native to Mexico. Hardy plant that has naturalized and is widely cultivated in the Indian subcontinent's gardens. Prefers a location in full sun. Can be invasive if left unchecked. **USES** Trained over fences and boundaries, it is a quick-growing vine that makes an attractive screen. Food plant for butterflies and suitable for planting in butterfly gardens. **ETYMOLOGY** Genus name is derived from the Greek *antigonon*, which means at opposite angles, referring to the angular stems. The epithet *leptopus* refers to the slender stems.

Garlic Pear Tree ■ *Crateva adansonii* subsp. *odora*

DESCRIPTION Small deciduous tree with a spreading crown. Grows to 8–10m in height. Leaves compound, with oval leaflets that are sharply tapered and arranged in groups of threes. Flowers attractive and appear when tree is nearly leafless in April–May. White petals fade to yellow; they surround numerous stamens and a single stalk of ovary. Attracts bees and other insects for pollination. **HABITAT** Native to India, Myanmar and Sri Lanka. Commonly found in the Deccan peninsula, West Bengal, Assam, Uttar Pradesh and Bihar.

USES Ornamental tree planted in Indian gardens and parks. Considered sacred by Islamic and Hindu faiths, and commonly found in historic monument complexes of Delhi such as Qutub Minar and Safdarjung's Tomb. Leaves, roots and bark of the root used in Ayurveda medicines. **ETYMOLOGY** Genus is named after Cratevas, a Greek botanist of the first century BC.

Camellia ▪ *Camellia japonica*

DESCRIPTION Handsome evergreen shrub with a dense, compact crown. Grows to 3–4m in height. Glossy green leaves oblong with tapered ends and finely serrated edges. They are simple and arranged alternately.

C. japonica is a predominant species from which most cultivars derive. Flowers are large and showy. They can be single or double, and of various shades of red, pink or white. They bloom in the cold months in November–March. **HABITAT** Native to China, Japan and Korea. Widely cultivated in cooler climate in the Indian subcontinent, such that of Northeast India and across the lower Himalayas. Thrives in well-drained, acidic soil, and prefers a position in partial shade. **USES** Ornamental shrub planted for its glossy foliage and strikingly beautiful flowers. **ETYMOLOGY** Genus is name in honour of George Joseph Kamel, a German missionary and naturalist in the Philippine isles. The epithet *japonica* refers to Japan, which is one of its native habitats.

Canna ■ *Canna indica*

DESCRIPTION Marginal aquatic plant with attractive foliage and flowers, growing in clumps from rhizomes to 1–1.5m in height. Large, broad leaves with sharply tapered tips rise upright, with leaf sheaths overlapping stem. Flowers are small, have slender petals and are borne on a tall spike. They are crimson red and sometimes yellow, and bloom

nearly all year round. The most common cannas are hybrids that have large flowers of various colours, including yellow, pink, orange, red and shades in between. Species such as C. *striata* have variegated leave.

HABITAT Native to tropical parts of America. Commonly cultivated in India, and can be seen growing wild along roadside drains and waterbodies. Thrives in moist soil conditions and positions in full sun. **USES** Hardy, ornamental plant grown in Indian parks and gardens. Also suited for planting in constructed wetlands and reedbeds, where grey water is filtered by root-zone treatment. **ETYMOLOGY** Genus name *Canna* is a Greek term for a reed. The epithet *indica* is a misnomer, as its origin is not India.

Canna hybrid varieties

Glory Lily ■ *Gloriosa superba*

DESCRIPTION Slender, graceful climber with deciduous leaves, growing from tuberous rhizomes to 2m tall. Lance-shaped leaves simple and arranged in opposite pairs. They have a curling tendril at the tip which helps the plant to latch on and climb. Flowers showy and attractive, at first appearing predominantly yellow and turning to red with time. Petals wavy and upturned, revealing prominent stamens. Flowers bloom in late spring to early autumn, and attract birds and butterflies. **HABITAT** Native to southern Africa and tropical Asia, including India in the Himalayan foothills, Tamil Nadu, Andra Pradesh and West Bengal. Widely cultivated in Indian gardens and requires a position in full sun, but tolerates partial shade. **USES** Ornamental shrub grown as a specimen plant in gardens. Used in traditional medicines such as Ayurveda for external applications to treat skin ailments and snakebite. Grown commercially in Tamil Nadu for its alkaloid extract. **ETYMOLOGY** Genus name means glorious, and epithet name *superba*, or superb, describes the splendid beauty of the flowers.

Egyptian Star Cluster ■ *Pentas lanceolata*

DESCRIPTION Evergreen shrub with a low spreading habit, growing to 0.5m in height. Leaves lance shaped with prominent veins and arranged in opposite pairs. Small flowers appear in dense, rounded clusters, and are star shaped with five petals and long, tubular

ends. Colours range from white, pink and mauve, to red, and flowers bloom nearly all year round. **HABITAT** Native to much of Africa and Yemen in southern Arabia. Cultivated widely across the plains of the Indian subcontinent, and flowers profusely as a perennial where the winters are not too severe. However, it is seasonal in places of extreme climate with hot summers and cold winters, such as Delhi. **USES** Ornamental plant used as a filler for garden beds. Also does well as a pot plant. Suitable for butterfly gardens. **ETYMOLOGY** Genus name is derived from the Greek *pente*, five, referring to the five petals of the flower, and the epithet *lanceolata* refers to the lance-shaped leaves.

Fire Bush ■ *Hamelia patens*

DESCRIPTION Medium-sized evergreen shrub with dense foliage that grows to 2m in height. Fine leaves oval shaped with tapered ends, and have a red tinge when new. Flame-orange flowers tubular, small and numerous, arranged in a cluster at the tips of their stalks. They bloom nearly year round and attract sunbirds, butterflies and bees. **HABITAT** Native to tropical America, from Florida to Central and South America. Cultivated in most parts of the Indian subcontinent and prefers a position in full sun. Also does well in

partial shade. **USES** Attractive foliage and flowers. Used as a filler plant for garden beds. Also often trimmed and maintained as a hedge. Suitable for butterfly gardens since it is a food plant for butterflies. All parts of the plant are used in herbal remedies, and known to be antibacterial, antifungal and anti-inflammatory. **ETYMOLOGY** Genus is named after the French botanist Henry Louise Hamel de Monceau, and the epithet *patens* refers to the spreading habit of the plant.

Gardenia ■ *Gardenia jasminoides*

DESCRIPTION Large evergreen shrub with a rounded crown. Grows to 3m in height. Glossy greepentasn leaves are oblong with tapered ends, and have prominent veins. Large white flowers have a sweet scent that becomes intense at night, attracting nocturnal moths for pollination. Single-flower varieties are funnel shaped, but double-flower varieties are more common. White petals fade to yellow as they age. Flowering occurs in summer to the cold winter months, in June–December. **HABITAT** Native to China and Japan. Known to have

been in cultivation there for nearly 1,000 years. Cultivated widely in most regions of India, and prefers a partially shaded location. Also grows well in full sun. **USES** Cultivated as an ornamental shrub for parks and gardens. Night-scented white blossoms make this a suitable plant for a moonlight garden. Floral oil used in perfume making and cosmetics. Also used for flavouring tea. **ETYMOLOGY** Genus is named after a Scottish physician and botanist, Dr Alexander Garden, and the epithet *jasminoides* means resembles jasmines.

Ixora ■ *Ixora coccinea*

DESCRIPTION Medium-sized evergreen shrub with glossy green leaves. Grows to 2–2.5m in height. Leaves simple, oval shaped, with tapered ends and arranged in opposite pairs. Scarlet-red flowers bloom nearly year round and appear profusely in rainy season. Yellow, pink and orange flower varieties are becoming popular. Small flowers salverform in shape, with a narrow tube opening out to a cross pattern of four petals. They are arranged in dense clusters. **HABITAT** Native to Western and South India, Sri Lanka and Bangladesh. Cultivated widely throughout India. **USES** Cultivated as an ornamental garden plant for its attractive flowers. Leaves, bark, roots and flowers used in medicines, and have anti-inflammatory, astringent and analgesic properties. **ETYMOLOGY** Genus name *Ixora* is derived from the Sanskrit *Isvara*, meaning Lord, referring to the god Shiva, to whom the flower is offered. The epithet *coccinea* refers to the scarlet colour of the flowers.

Small Flowered Ixora ■ *Ixora parviflora*

DESCRIPTION Large evergreen shrub with a compact crown. Can grow to 3–4m tall. Glossy green leaves oblong, simple and arranged in opposite pairs. White flowers appear in large, round clusters in March–April and are sweetly scented. They are small and salverform, with four flat-faced petals and a tubular end. **HABITAT** Native to India,

Bangladesh and Sri Lanka. Thrives in a hot, humid climate in a location in full sun or partial shade. Found in South India, the Western Ghats, Bihar and Orissa. Cultivated widely throughout India. **USES** Ornamental shrub grown for its fragrant flowers. Important medicinal plant locally known as *Nevari*. All parts of the plant are used in traditional medicine remedies in Ayurveda and Siddha systems. **ETYMOLOGY** Genus name *Ixora* is derived from the Sanskrit *Isvara*, meaning Lord, referring to the god Shiva.

Panama Rose ■ *Rondoletia odorata*

DESCRIPTION Evergreen woody shrub that can grow to 2m in height. Dark green leaves

oblong with a thick, leathery texture, and arranged in opposite pairs. Flowers small, tubular in shape and open out to flat-faced petals. Orange flowers have deep yellow centres, and despite the name are not scented. They bloom in warm summer months up to autumn. **HABITAT** Native to Panama and Cuba. Cultivated widely in gardens of the Indian subcontinent. Thrives in full sun, and also does well and flowers in semi-shade. **USES** Ornamental plant grown in gardens for its foliage and flowers. Host plant for butterflies, and used for planting in butterfly gardens. **ETYMOLOGY** Genus is named after French physician and biologist Guillame Rondelet, and the epithet name *odorata* refers to the roots, which are fragrant.

Red Flag Bush ■ *Mussaenda erythrophylla*

DESCRIPTION Large shrub with dense foliage and scrambling habit. Grows to 3m in height. Can be pruned and maintained in a compact form. Floral configuration is curious, with small, pale yellow flowers surrounded by bright red sepals that make the inconspicuous flowers visible to their pollinators. Rich source of nectar and attracts butterflies.

HABITAT Native to West Africa. Cultivated in hot, humid regions of South India and prefers a moist climate.

USES Ornamental garden plant for parks and gardens. Host plant for butterflies and ideal for creating butterfly gardens.

ETYMOLOGY Genus name is derived from the Singhalese *mussende*, a term for a species of this genus. The epithet *erythrophylla* refers to the red modified leaves or sepals.

Crepe Ginger ■ *Cheilocostus speciosus*

DESCRIPTION Evergreen herbaceous shrub that grows to 2m in height from rhizomes. Leaves lance shaped and arranged in spirals along stem. White flowers have a crepe-like texture, hence the common name. They bloom in July–September, appearing on top of a spike of red overlapping bracts. **HABITAT** Native to tropical Asia, including the Indian

subcontinent. Plant of subtropical and tropical climates, and prefers humid conditions. Grows in the wild in the Himalayan foothills, the Western Ghats, Northeast India and Orissa. **USES** Cultivated in gardens for its ornamental foliage and flowers. Host plant for butterflies, and suited for butterfly gardens. Rhizomes and leaves are used in Ayurvedic medicines, where it considered a tonic, astringent, purgative and aphrodisiac. **ETYMOLOGY** Genus name is derived from the Greek *cheilos*, lip, and *costus*, which was the genus in which it was previously classified. The epithet *speciosus* means beautiful or striking.

Red Button Ginger ■ *Costus woodsonii*

DESCRIPTION Evergreen rhizomatous shrub with attractive foliage. Grows to 1.5–2m in height. Large leaves oblong and pointed with leaf sheaths overlapping stems. Small orange flowers tubular and borne on a red-coloured spike of overlapping bracts. A single flower appears on each spike in a day. Flowers bloom in March–September, and attract bees and sunbirds for pollination. **HABITAT** Native to Nicaragua and Colombia in South America. Commonly cultivated in hot, humid climate of South India. **USES** Ornamental garden plant used as a filler in planter beds. Long-lasting bracts make good cut flowers. **ETYMOLOGY** Genus name *Costus* is a Latin term for fragrant plants. Species is named after Dr Robert Woodson, a curator of the Missouri Botanical Garden herbarium.

Rangoon Creeper ■ *Combretum indicum*

DESCRIPTION Woody climber with vigorous growth that reaches to 6m in height. Bright green leaves oblong with tapered ends, and arranged in opposite pairs. Blooms March–November, in attractive drooping clusters. Flowers salverform, with long tubes opening flat into five petals. They are white when the flowers open at night and turn pink in the day, slowly changing to red. Rich in nectar and intensely fragrant at night, they attract bees

and sunbirds in the day, and nocturnal moths at night. **HABITAT** Native to East Africa, Southeast Asia and Malaysia. Hardy plant, widely cultivated throughout India. Prefers a location in full sun but also thrives in semi-shade. **USES** Ornamental climber grown for its profuse fragrant flowers. Suited for planting in moonlight gardens for its scent, which is most intense at night. **ETYMOLOGY** Genus name, earlier known as *Quisqualis* and now changed to *Combretum*, refers to a term given by Pliny, the Roman philosopher and naturalist, to a climbing plant. The epithet *indicum* refers to India, its place of origin.

Calendula ▪ *Calendula officinalis*

DESCRIPTION Bushy annual shrub with dense foliage. Small plant that grows to 0.3–0.5m high. Aromatic leaves oblong and bright green. Flowers disc shaped with petals radiating centrally around a dense centre portion of florets. Colours vary from yellow to orange.

HABITAT Native to southern Europe in Mediterranean region, and commonly grown thoughout India as a winter seasonal plant. **USES** Popular as a pot plant and for filling colourful flower beds in gardens. Petals edible and added to salads and used as a garnish. Can also be brewed to make tea. Bitter leaves rich in minerals and vitamins, and their extracts are used medicinally. **ETYMOLOGY** Genus name is derived from the term *calendae* and refers to the first day of the month.

Chrysanthemum ■ *Chrysanthemum* spp.

DESCRIPTION Herbaceous shrub comprising annual and perennial plants with profuse ornamental flowers. Leaves deeply lobed and covered in fine hairs. Genus includes more than 50 species, classified by the varying shapes of the flowers – single, pompon or double, spoon type, spider form, quill shaped, anemone centred, intermediate, reflexed and incurved. Each flower head comprises numerous tiny florets. Numerous flower colours, including yellow, white, red, purple and orange. In North Indian plains, flowers appear in

early winter, while in mountains plants bloom in summer. **HABITAT** Native to Asia and northeastern Europe. Many species originated in China, where they were known to be in cultivation as early as the fifteenth century BC. **USES** Cultivated as an ornamental seasonal plant in Indian gardens. Blooms long lasting and popular as cut flowers. Numerous species and cultivars are available, and there are many specialist nurseries in cities such as Kolkata, Pune and Dehradun in India. Extracts from flowers used as a natural insecticide. Plant is now known to have air-cleaning properties, according to NASA. **ETYMOLOGY** Genus name is derived from the Greek *chrisos*, gold, and *anthemon*, flower.

Chrysanthemum varieties

Dahlia ■ *Dahlia pinnata*

DESCRIPTION Bushy, herbaceous and tuberous shrub that grows tall and upright to 0.75–1.5m in height. Bright green leaves deeply lobed. Large flowers showy, disc shaped and appear in winter. Dahlias are classified into groups according to flower type – single, pompon or double, anemone, collerette, water lily, decorative, ball, cactus, semi-cactus, funbriated and orchid. Flower colours include white, pink, yellow, mauve and red. *The*

Encyclopaedia of Dahlias by Bill McLaren lists nearly 700 varieties of Dahlia. **HABITAT** Native to Mexico, where it is the national flower. Introduced in India, and widely cultivated as a winter seasonal plant in the plains, and as a summer seasonal in the mountains. Suited to a location in full sun. **USES** Cultivated as an ornamental garden plant for its showy and decorative flowers. **ETYMOLOGY** Genus is named after the Swedish botanist Andreas Dahl.

Dahlia hybrid varieties

Gazania ■ *Gazania rigens*

DESCRIPTION Herbaceous perennial often grown as an annual. It has a clump-forming habit and grows to 0.3m in height. Leaves long, narrow and deeply lobed, with undersides that are silver coloured due to the presence of fine hairs. Daisy-like flowers showy and bloom in cold months in the plains, December–March. They are borne solitarily on tall stalks that rise above the foliage. Yellow-orange ray petals surround a central disc of florets. Bicoloured flowers are seen in hybrid cultivars of *G. rigens*. Flowers open fully in the sun. **HABITAT** Native to South Africa. Widely cultivated in India as a perennial in the hills and as a seasonal plant in the plains. **USES** Commonly used as a bedding or edging plant in gardens, and grown for its attractive foliage and flowers. Popular as a cut flower. **ETYMOLOGY** Genus is named after the Greek scholar Theodore de Gaza, who translated many important botanical works. The epithet *rigens* means stiff or rigid.

Marigold ■ *Tagetes erecta*

DESCRIPTION Medium-sized annual shrub that can vary in height at 0.1–2m. Dark green leaves serrated and aromatic. Flowers profuse and appear in autumn and throughout winter. They are disc shaped and have large petals in the periphery and dense florets in the centre. Colours range from yellow and orange to deep maroon, and sometimes are a combination of maroon and yellow. **HABITAT** Native to southern parts of North America – Guatemala and Mexico – and widely cultivated in India. Suited to both clay and sandy soils, and needs full sun. **USES** Cultivated on large scale as a crop for its ornamental flowers, which are made into garlands and used in ceremonies as decoration and votive offerings. Also grown in gardens and parks as a seasonal plant, and known to repel mosquitoes and other insects. Petals edible, and a dye is obtained from them, which is used as a colour for foods such as pasta, mayonnaise and dairy items. Marigolds are used in butterfly garden planting. Leaves, flowers and roots are known to have medicinal properties, and are used to treat stomach ailments. **ETYMOLOGY** Genus name is derived from *Tages*, an Etruscan deity who was considered to be the god of the underworld. The epithet *erecta* describes the plant's upright growing habit.

Sunflower ■ *Helianthus annus*

DESCRIPTION Genus of nearly 70 species of herbaceous annual shrub. Grows tall and upright to a height of 3.0m. Leaves heart shaped with fine bristling hairs. Bright yellow flowers disc shaped with a dark brown centre that contains numerous small florets. They appear in the warm summer months up to autumn. Flowers face the direction

of the sun, following the movement of the sun from east to west during the day. **HABITAT** Native to Central America. Widely cultivated in India as a cash crop and as a seasonal flowering plant in gardens. **USES** While Sunflowers are grown for their ornamental value, they are primarily cultivated as a crop for the cooking oil extracted from the seeds. These are also eaten, raw or roasted. **ETYMOLOGY** Genus name in Greek, *Helios*, means sun, arriving at the common name Sunflower. The ancient civilizations of Peru (Incas) worshipped the plant as the symbol of the sun.

Wedelia ■ *Sphagneticola trilobata*

DESCRIPTION Evergreen shrub with creeping growth habit, spreading along the ground. Branches usually less than 0.5m long. Glossy green leaves dense and deeply lobed. Yellow flowers disc shaped like asters, and attract insect pollinators such as bees and butterflies. Flowers appear sporadically throughout the year. **HABITAT** Native to Central and South

America. Natural habitats are marshes and beaches. Cultivated widely in India and does well in both sunny and partially shaded locations. **USES** Used as a groundcover and filler in plant beds. Has vigorous growth and is invasive if left unchecked. **ETYMOLOGY** Genus name is derived from *sphagnum*, referring to the plant's ability to survive in waterlogged areas, and the epithet *trilobata* describes its thrice-lobed leaves.

Crown Flower ■ *Calotropis procera*

DESCRIPTION Evergreen shrub with an upright habit. Grows to 2–3m in height. Grey-green stems and foliage. Oblong leaves arranged in opposite pairs. Flowers have a waxy texture and upstanding petals like a crown, hence the common name. Petals creamy-white with tips tinged with mauve, whereas those of the India Rubber Vine (see p. 000) are creamy white or mauve, and the petals open out instead of standing up. Free flowering and blooms all year round. Milky sap is toxic.

HABITAT Native to the Indian subcontinent, south China, Malaysia and Indonesia. Widely distributed in India and grows wild in wasteland and roadsides. Tolerates drought and saline soil conditions. **USES** Often grown near temples, and the flowers are used as votive offerings and in the worship of Lord Shiva by Hindus. Important medicinal plant known locally as *Arka*. All parts of the plant are used in traditional Ayurveda and Unani practices, and it is known to be antiseptic, emetic and purgative. **ETYMOLOGY** Genus name derives from the Greek *kalos*, beautiful, and *tropos*, boat, referring to the beautiful bowl-shaped flowers. The epithet *procera* means tall or upright, which is the habit of the plant.

Calotropis gigantea

Crape Jasmine ■ *Tabernaemontana divaricata* 'Flore Pleno'

DESCRIPTION Evergreen shrub, multi-branched with a spreading crown. Grows to 2m in height. Dark green leaves simple, lance shaped and arranged in spirals along branches. White flowers salverform in shape, with a thin tube opening out to five petals like a pinwheel. *T. divaricata* 'Flore Pleno' is a double-flower cultivar that is fragrant, especially

at night, and attracts nocturnal moths. Blooms in summer and is most profuse during monsoons. **HABITAT** Native to India, Myanmar and Thailand. Cultivated widely in gardens throughout India. **USES** Popular ornamental shrub in Indian gardens and parks. Suited for planting in moonlight gardens for its white scented flowers. **ETYMOLOGY** Genus is named after Jacobus Theodorus von Bergzabern, a German herballist. Bergzabern translates to mountain tavern or *tabernaemontanus* in Latin. The epithet *divaricata* means wide spreading, which is the growth habit of the plant.

Desert Rose ■ *Adenium obesum*

DESCRIPTION Succulent shrub with a bulbous, fleshy trunk. Grows to 2m in height and looks like a miniature tree or bonsai. Deciduous leaves oblong and arranged in whorls along branches. Large, trumpet-shaped flowers are showy, and colours range from red to pink with white inner throats; sometimes bicoloured. **HABITAT** Native to tropical West Africa and Arabian Peninsula. Natural habitat is arid and semi-arid. Cultivated widely in India as a garden ornamental, and thrives in sandy, well-drained soil. **USES** Ornamental plant grown for its showy flowers and form. Suited for planting in rock gardens or as a container plant. An important medicinal plant in Africa. **ETYMOLOGY** Genus name is derived from Aden in Arabia, where the plant was first documented. The epithet *obesum* refers to the fat or fleshy stem.

Easter Lily Vine ▪ *Beaumontia grandiflora*

DESCRIPTION Large, woody climber that can grow to 10m in height. Evergreen leaves simple, oblong in shape with a hairy underside, and arranged in opposite pairs. White flowers large and fragrant, borne in clusters. They are trumpet shaped and appear in late spring, in March–April. Attracts insect pollinators such as bees and butterflies. **HABITAT** Native to east Asia, eastern Himalayas in India, Nepal, Bangladesh, Myanmar, and south

China up to Vietnam. Prefers a sunny location and adapts well to hot, dry climate conditions. **USES** Ornamental climber grown as a specimen plant in gardens. Leaves and roots used in treating wounds and fractures. **ETYMOLOGY** Genus is named after Mrs Diana Beaumont of Bretton Hall, England, who was a celebrated patron of horticulture, and the epithet *grandiflora* means large flowered.

Frangipani ▪ *Plumeria obtusa*

DESCRIPTION Small, multi-branched evergreen tree with a rounded crown. Grows to 6m in height. Leaves oblong with prominent veins and thick, leathery texture. Tips of leaves blunt and rounded, differentiating it from other species. Flowers funnel shaped with rounded, waxy petals. Although it has no nectar, its fragrance at night attracts nocturnal moths. Creamy-white flowers with a tinge of yellow in the centre. **HABITAT** Native to Central and South America, from Florida and Mexico, to Guatemala and Venezuela. Widely cultivated in the Indian subcontinent. **USES** Ornamental in form, foliage and flowers, and a popular plant in Indian gardens and parks. Often planted near temples, and flowers are used as votive offerings. Essential oil from flowers used in perfumes and cosmetics. **ETYMOLOGY** Genus is named after French monk and botanist Charles Plumier. The epithet *obtusa* means blunt, referring to the leaves.

Frangipani ▪ *Plumeria rubra*

DESCRIPTION Small deciduous tree with a multi-branching crown. Grows to 6m in height. Leaves oblong with prominent veins and thick, leathery texture. Tips of leaves tapered. Flowers creamy-white, pink or red, and sometimes bicoloured. They are fragrant

and funnel shaped, with waxy petals that have tapered tips. Also known as *P. acuminata*. **HABITAT** Native to Central and South America, from Mexico to Venezuela. Widely cultivated in the Indian subcontinent and thrives in a tropical climate. **USES** Popular in parks and gardens, and often planted near temples for its flowers, which are used as votive offerings. Essential oil from flowers used in perfumes and cosmetics. Medicinally, sap is used to treat skin ailments. **ETYMOLOGY** Genus is named after French monk and botanist Charles Plumier. The epithet *rubra* refers to some varieties of the species that are red in colour.

Fiddle Leaf Plumeria ▪ *Plumeria pudica*

DESCRIPTION Small tree with bushy, rounded crown. Grows to 4m in height. Leaves long and narrow, flaring at tip to form a triangular point that distinguishes this species. They have a thick, leathery texture and prominent veins, and are arranged in whorls. White flowers funnel shaped with waxy petals that are broad and sharply tapered. **HABITAT** Native to Central and South America, from Panama to Venezuela. New

introduction to the Indian landscape. **USES** Cultivated as a specimen plant in gardens for its attractive foliage and flowers. Also thrives as a pot plant. **ETYMOLOGY** Genus is named after French monk and botanist Charles Plumier. The epithet name is derived from the Latin *pudicus*, pure and chaste, referring to the flowers, which are often used in bridal bouquets.

Golden Trumpet Vine ■ *Allamanda cathartica*

DESCRIPTION Evergreen shrub with scandent, rambling growth. Can be trained as a climber and grows to 5–6m in height. Broad leaves have tapered ends and are arranged in whorls of four leaves. Yellow flowers large and trumpet shaped, appearing throughout summer. All parts of the plant, including the sap, are toxic. **HABITAT** Native to Brazil in South America, and naturalized in tropical and subtropical parts of the world. In India, widely cultivated as a popular garden plant. Prefers a location in full sun. **USES** Ornamental plant that is commonly trained as a climber on arbours and trellises. Leaves and roots used in traditional herbal medicines as a purgative. **ETYMOLOGY** Genus is named after Dr Frederic-Louis Allamand, a Swiss botanist, and the epithet *cathartica* refers to the leaves, which have cathartic, purgative properties.

Greater Periwinkle ■ *Vinca major*

DESCRIPTION Evergreen shrub with glossy green foliage and a low, spreading habit. Can grow to 0.5m in height. Oblong leaves have pointed tips and are arranged in opposite pairs. A variegated leaf version has attractive foliage and flowers. Flowers purple or white and sometimes bicoloured; they are salverform in shape with a tubular end opening out to five flat-faced petals. They bloom from spring to autumn, and attract bees and moths for pollination. **HABITAT** Native to southern Europe and east Mediterranean – Anatolia and

North Africa. Thrives in subtropical regions of the Indian subcontinent. **USES** Hardy ornamental shrub used as a groundcover in garden beds. Also used in medicines, as it contains alkaloids that are a stimulant. **ETYMOLOGY** Genus name is derived from the Greek *vincere*, to bind, referring to the long, creeping stems used to make wreaths. The epithet *major* refers to the large leaves that differentiate it from the small-leaved *V. minor*.

India Rubber Vine ■ *Cryptostegia grandiflora*

DESCRIPTION Vigorous evergreen climber that can grow to 10m in height, sometimes covering entire tree canopies. Dark green leaves elliptical in shape, and arranged in opposite pairs. Trumpet-shaped flowers mauve-purple and white internally, and appear

in summer. Pollinated by bees and other insects. All parts of the plant are toxic. **HABITAT** Native to Madagascar. Widely cultivated in India and thrives in semi-arid climate conditions. **USES** Ornamental climber trained over fences and arbours for its attractive flower clusters. Milky sap or latex is used to make rubber of a quality comparable to that of the *Hevea* tree. Considered an invasive species in Australia.

Malati ■ *Aganosma heynei*

DESCRIPTION Large evergreen climber that grows to 10m in height. Dark green leaves elliptical with prominent veins, and arranged in opposite pairs. White flowers highly fragrant and bloom in summer, especially during monsoons in June–September. Star-shaped flowers have five petals that are slightly twisted and attract bees. **HABITAT** Native to India. Commonly planted in Indian gardens and known as *Malati*. Thrives in partially shaded locations and prefers dry soil conditions. **USES** Ornamental climber grown for its fragrant flowers. Also used in traditional Indian medicine as an emetic and anti-parasitic.

Poison Arrow Plant ■ *Acokanthera oblongifolia*

DESCRIPTION Evergreen shrub with a wide, spreading crown. Grows to 3m in height. Dark green leaves waxy and oblong in shape, arranged in opposite pairs. Flowers bloom in spring, in April–May. Small white flowers tubular, packed in dense clusters and intensely fragrant. They attract butterflies. Fruits berry-like and dark purple in colour, and toxic when unripe. **HABITAT** Native to western region of South Africa; natural habitat is in coastal dune forests. Cultivated in India, and an understorey plant growing under large trees and preferring a partially shaded position. **USES** Ornamental plant grown for its fragrant flowers. In its native region, used in traditional medicines to treat snake and insect bites. Toxic stem, bark and roots are used to make poison for arrow tips. **ETYMOLOGY** Genus name in Greek refers to the sharp-pointed anthers of the flower. The epithet *oblongifolia* refers to the oblong leaves.

Oleander ■ *Nerium oleander*

DESCRIPTION Evergreen shrub with bushy growth and rounded form. Can reach to 3m in height. Leaves narrow and lance shaped, arranged in groups of three leaves. Trumpet-shaped flowers appear in clusters at ends of branches nearly all year round. Most double-flowering cultivars are fragrant and attract insect pollinators by their scent, since they do not produce nectar. All parts of the plant are toxic. **HABITAT** Native to the Mediterranean region, Iran, Indian subcontinent and south China. Hardy plant that

is cultivated throughout India and prefers a location in full sun. **USES** Ornamental shrub commonly planted along roadsides and in parks, since it is not eaten by cattle due to its toxic nature. Roots and leaves contain oleandrin compound, which is used in traditional medicines. **ETYMOLOGY** Genus name is derived from the Greek *nerion*, moist, since it prefers moist conditions. The epithet *oleander* refers to the long, thin leaves that resemble the leaves of Olive trees.

Lady's Eardrop ■ *Fuchsia* spp.

DESCRIPTION Evergreen shrub with delicate form and foliage; some species have a drooping form, while others grow upright, to 2m in height. Leaves elliptical with sharp, pointed ends and arranged in opposite pairs. Flowers appear in drooping clusters and have an interesting structure, resembling a lady's drop earrings, hence the common name.

Inner petals mostly purple and have an outer layer of sepals that vary in colour from white, pink and mauve, to red. They are pollinated by insects such as bees, and in their native habitat by hummingbirds. **HABITAT** Native to tropical and subtropical parts of Central and South America. In India popularly cultivated in cooler climates of hill stations of Northeast India. Prefers a cool, moist climate and a location in partial shade. **USES** Ornamental plant cultivated in gardens, and especially popular as a pot plant. Suited for planting in hanging baskets due to its droopy form. **ETYMOLOGY** Genus is named after Leonhart Fuchs, a sixteenth-century German physician and botanist.

Barometer Bush ■ *Leucophyllum frutescens*

DESCRIPTION Woody shrub with semi-deciduous foliage. Grows to 2.5m in height. Attractive foliage comprises silver-grey leaves, oblong in shape and arranged alternately. Mauve-purple flowers bell shaped. They are borne solitarily and attract bees, butterflies

and birds. They bloom in warm summer months and are most profuse in monsoons, since their flowering is triggered by moist soil conditions and high humidity. **HABITAT** Native to Texas, USA, and Mexico, and widely cultivated in India. Hardy plant that is drought and heat tolerant, favouring a location in full sun. **USES** Ornamental shrub grown for its attractive foliage and flowers. Also suitable for planting in desert gardens. **ETYMOLOGY** Genus name *Leucophyllum*, or white leaves, refers to its silver-grey colour. The epithet *frutescens* refers to its growth habit, which can become bushy or shrubby.

Firecracker Plant ■ *Russelia equisetiformis*

DESCRIPTION Evergreen shrub with drooping form and low, spreading habit. Branches slim and bare with no leaves. Red flowers tubular in shape and appear in drooping clusters almost throughout the year, and attract bees and butterflies. **HABITAT** Tropical plant native to Mexico. Cultivated widely in the plains and low altitudes of the Indian subcontinent. Resistant to drought and saline conditions, it is a hardy plant that adapts

to most climatic conditions. **USES** Popular in gardens as a groundcover. Known to be a food plant for butterflies, and can be planted in butterfly gardens. Cultivated in Central America as a medicinal plant, and known to have antibacterial and anti-inflammatory properties. **ETYMOLOGY** Genus is named after Scottish naturalist Dr Alexander Russell. The epithet *equisetiformis* refers to the form, which resembles a horsetail.

Bougainvillea ▪ *Bougainvillea* spp.

DESCRIPTION Large, woody climber with spiny branches that climbs to 5–6m in height.
Leaves broadly oval with tapered ends and arranged alternately. Colourful bracts that look like petals surround true flowers that are small and tubular, grouped in triangles of three. Bracts can be white, pink, red, orange or mauve. The conspicuous bracts and nectar of the tiny flowers attract insect pollinators such as bees and moths. Blooms nearly throughout the year, and profuse flowering is triggered by dry soil conditions. **HABITAT** Native to tropical South America, including Brazil, Peru, Colombia and Ecuador. Widely cultivated throughout the Indian subcontinent. **USES** Numerous hybrid cultivars have been developed, and this is a popular ornamental plant in gardens of the region. Used as a hedge plant along road medians, as a climber along fences and boundary walls, and even as a container houseplant. **ETYMOLOGY** Genus is named after French navigator L. A. de Bougainville, who collected the species from Rio de Janeiro on his trip around the world. Main species from which cultivars have been developed are *B. spectabilis*, *B. peruviana*, *B. glabra* and *B. buttiana*.

Four O'Clock Plant ■ *Mirabilis jalapa*

DESCRIPTION Evergreen shrub that spreads low over the ground and can grow to 1m tall if supported. Oblong leaves have acute pointed tips and are arranged in opposite pairs. Bright pink flowers trumpet shaped and borne in clusters, and bloom throughout the year. Buds open in the evening at 4–8 p.m., hence the common name. Flowers fragrant at night but without any scent in the day; they attract nocturnal moths for pollination. **HABITAT**

Native to South America, and naturalized and grows wild in India. Cultivated as a hardy garden plant, and prone to being invasive if left unchecked. **USES** Quick growing and always in bloom, this plant is suitable for use as a groundcover in gardens. Known to be a butterfly food plant, it is also suited for planting in butterfly gardens. Roots used medicinally as a purgative. **ETYMOLOGY** The name *Mirabilis jalapa* was given by the Swedish botanist Carl Linnaeus. *Mirabilis* means wonderful, referring to its colours, and the epithet *jalapa* probably refers to its origin in Jalapa in Guatemala.

Garden Geranium ■ *Pelargonium* x *hortorum*

DESCRIPTION Small shrub that is low growing and has a rounded form. Grows to 0.5m in height. Leaves rounded, with wavy margins and dark brown ring in centre. Flowers appear in attractive clusters and range in colour from white, to pink, orange and red. They attract bees and butterflies. **HABITAT** Also known as Zonal Pelargonium, this is a hybrid of *P. zonale* and *P. inquinans*. Parent species are native to South Africa. Widely cultivated in cooler climes of the Indian subcontinent, and grown as a seasonal winter plant in hot, dry regions. **USES** Ornamental shrub, popular as a houseplant and thriving as a container plant. **ETYMOLOGY** Genus name is derived from the Greek *pelargos*, stork, referring to the fruit, which has a beak like a stork. The epithet *hortorum* is drawn from the Latin *hortus*, garden.

Scented Geranium ■ *Pelargonium graveolens*

DESCRIPTION Bushy evergreen shrub with rounded form. Grows to 0.5m in height. Simple leaves strongly aromatic and deeply lobed, with velvet-like texture due to fine hair covering. Flowers small and delicate, and white or lilac in colour. They bloom in late spring, in March–May, and appear in small clusters. **HABITAT** Native to South

Africa. Cultivated in cooler climes of the Indian subcontinent at medium altitudes, as well as plains (Pune, Bangalore, Northeast India) with mild climatic conditions. **USES** Ornamental shrub grown for its aromatic leaves. Essential oils from leaves used in cosmetics. **ETYMOLOGY** Genus name is derived from the Greek *pelargos*, referring to the fruit, which is shaped like the beak of a stork. The epithet *graveolens* means heavily scented, referring to the rose-like smell of the leaves.

Wallich Geranium ■ *Geranium wallichianum*

DESCRIPTION Low-height shrub with a habit of trailing on the ground. Grows to 0.3m tall. Leaves deeply lobed into three or five segments and have serrated edges. Pink-purple flowers delicate and cup shaped, comprising five petals that have dark line markings. They bloom in summer, in June–September, and are borne solitarily. **HABITAT** Native to the Himalayan range from Afghanistan to Bhutan, this is a plant of temperate climates. Grows wild along

mountain slopes and valleys. Best seen in the Valley of Flowers in the Indian state of Uttarakhand. Thrives in partial shade as well as full sun. **USES** Attractive foliage and flowers, which can be used in creating wildflower gardens. Roots and leaves used in traditional medicines. **ETYMOLOGY** Genus name is derived from the Greek *geranos*, crane, referring to the fruit, which resembles a crane's beak. The epithet name is in honour of Nathaniel Wallich, a Danish surgeon and botanist who was instrumental in the development of the Royal Botanic Garden in Calcutta.

Shell Ginger ■ *Alpinia zerumbet*

DESCRIPTION Evergreen shrub with attractive foliage. Grows to 2m in height, from rhizomes. Aromatic leaves are long and lance shaped, with leaf sheaths overlapping stem.

Flowers bloom in summer, especially during monsoons, in a drooping spike of pearl-white buds that open to reveal yellow-orange inner petals with markings to guide insect pollinators such as bees. **HABITAT** Native to tropical and subtropical Asia, including the Indian subcontinent. Grows along water courses, streams and shady slopes. **USES** Ornamental shrub grown for its showy foliage and flowers. Can be grown in reedbeds or constructed wetlands used to treat grey water by root-zone treatments. Rhizomes used medicinally to treat digestive disorders and skin ailments. Essential oil from flowers and rhizomes used as a culinary spice. Tea from fresh leaves is known to be enjoyed in Japan. **ETYMOLOGY** Genus was named after Italian botanist Prospero Alpini by Carl Linnaeus.

Turmeric ▪ *Curcuma psuedomontana*

DESCRIPTION Herbaceous shrub growing from rhizomes to 1m in height. Aromatic leaves long and lance shaped, and arranged alternately. Flowers small and inconspicuous,

appearing on a tall spike of overlapping bracts and opening one at a time. Bracts showy and colourful, and attract pollinating insects. **HABITAT** Most species of *Curcuma* are known to be from Southeast Asia – India, China and Vietnam. However, *C. longa* is cultivated and does not grow wild. **USES** Used as a culinary spice in cooking. An important medicinal plant in Ayurveda and Unani medicines. Rhizomes are boiled and dried before use, and the plant is known to be antifungal, antibacterial, antioxidant and anti-inflammatory. Yellow colour derived from rhizomes is used in cosmetics and to dye fabric. Considered auspicious in India, it is used in rituals and by Hindu brides before a wedding. **ETYMOLOGY** Genus name is derived from the Arabic *kumkum*, for its saffron-like colour.

Orange Ginger Lily ▪ *Hedychium coccineum*

DESCRIPTION Aromatic herb with upright form and clump-forming habit, rising from rhizomes. Grows to 2.5m in height. Long leaves lance shaped and arranged alternately.

Tall spike of orange-red blossoms arranged in tapering cluster. Flowers bloom through warm summer months. **HABITAT** Native to the Himalayan region of East India and Nepal up to south China. Found along edges of streams. Thrives in moist soil conditions and prefers a position in partial shade. **USES** Ornamental plant cultivated for its attractive flowers, which are used as votive offerings. **ETYMOLOGY** Genus name originates from the Greek *hedys*, snow, and *chios*, sweet, in reference to the sweet smell of the flowers. The epithet *coccineum* refers to the red colour of the flowers.

White Ginger Lily ■ *Hedychium coronarium*

DESCRIPTION Aromatic herb with upright form and clump-forming habit, rising from rhizomes. Grows to 1.5m in height. Long leaves lance shaped and arranged alternately. Delicate white flowers borne in clusters and bloom throughout summer up to autumn. They are highly scented, especially at night, and attract nocturnal moths. Cultivars like *H. coronarium* var. *chrysoleucum* have a yellow centre among white petals. **HABITAT** Native to the Himalayan region of East India and Nepal up to south China. Found along edges of streams and in waterlogged areas. **USES** Ornamental plant cultivated for its fragrant flowers, which are used as votive offerings and sometimes worn by women to adorn their hair. Its essential oil is used in perfumes and cosmetics. Stems and rhizomes used in medicines, and known to be anti-inflammatory and diuretic. **ETYMOLOGY** Genus name originates from the Greek *hedys*, snow, and *chios*, sweet, in reference to the sweet smell of the white flowers. The epithet is derived from the word *corona* for crown, as the flowers are strung to make garlands.

Azalea ■ *Rhododendron yedoense*

DESCRIPTION Attractive semi-deciduous shrub with a spreading crown. Grows to 2m tall. Leaves lance shaped and arranged in whorls at branch ends. Trumpet-shaped

flowers showy and occur in shades of pink-purple. They bloom in spring, in April–May. Thousands of cultivars have been developed over centuries. **HABITAT** Native to Korea and Japan, and popularly cultivated in cooler climes of the Indian subcontinent. Natural habitat is in alpine woodland as an understorey plant. Prefers a location in partial shade. **USES** Azalea varieties are popular ornamental shrubs in gardens. They also thrive as pot plants. **ETYMOLOGY** Genus name is derived from the Greek *rhodon*, rose, and *dendrum*, tree. The epithet name is derived from the medieval Japanese name *Edo*, which is now Tokyo, referring to one of its places of origin.

Azalea hybrid varieties

Tree Rhododendron ■ *Rhododendron arboreum*

DESCRIPTION Small evergreen tree with a dense, rounded crown. Can grow to 5–6m in height. Glossy green leaves simple, elliptical in shape and arranged in whorls at ends of branches. Flowers bloom profusely in March–April and also in June–September, and are predominantly red in colour but sometimes pale pink. They are campanulate or bell shaped, and packed in tight clusters. **HABITAT** Native to the Himalayan range and southern Tibet. Plants of temperate climate, and prefer shaded, dappled sunlight. **USES** Popularly known as *Burans*, plants are used as specimen trees in gardens at high altitudes. Rhododendron juice is produced from the flowers. Leaves and flowers used in traditional medicines such as Ayurveda and homeopathy. **ETYMOLOGY** Genus name is derived from the Greek *rhodon*, rose, and *dendrum*, tree; the epithet *arboreum* also refers to its tree-like habit.

Rhododendron varieties

Lobster Claw ■ *Heliconia rostrata*

DESCRIPTION Tall evergreen shrub with clump-forming habit, growing to 2.5m in height, from rhizomes. Leaves lance shaped and simple, and arranged alternately. Flowers attractive and unusual, appearing in long, drooping spikes of claw-shaped bracts that are bright red in colour, from which small yellow flowers arise. In their native region, they are known to be pollinated by hummingbirds. **HABITAT** Native to South America, from Peru to Argentina, and thrives in tropical conditions in partial shade. Cultivated widely in the Indian subcontinent as a garden plant. **USES** Ornamental shrub grown for its showy floral arrangement. Bracts are long lasting and it is popular as a cut flower. **ETYMOLOGY** Genus name is derived from the Greek *helikonios*, derived from Mount Helicon in Greek mythology. The epithet *rostrata* means long, sharp point or tip, referring to the claw-shaped bracts.

Honeysuckle ■ *Lonicera japonica*

DESCRIPTION Vigorous climber with slender form and semi-deciduous foliage, which grows to 5m tall. Deep green leaves elliptical, with pointed ends and leathery texture. They are arranged in opposite pairs. Sweetly fragrant flowers bloom in early summer, in March–May. They are creamy-white with a long, tubular shape that opens out as two lips. Flower buds open at dusk and are pollinated by nocturnal moths. **HABITAT** Native to Korea and Japan, and popularly cultivated as a garden plant in cooler regions of the Indian subcontinent. **USES** Ornamental climber cultivated for its sweet-smelling flowers. Stem and flowers used in traditional medicines, and known to be antibacterial and antiviral. **ETYMOLOGY** Genus is named after German naturalist Adam Lonitzer; the epithet *japonica* refers to its region of origin, Japan.

Hardy Hydrangea ■ *Hydrangea macrophylla*

DESCRIPTION Deciduous shrub with dense foliage and rounded form. Grows to 2m tall. Dark green leaves oblong with acute pointed tips, and arranged in opposite pairs. Flowers bloom in summer, in large, rounded clusters comprising tightly packed bracts topped by minute flowers. Colours vary according to soil conditions; alkaline soil produces pink flowers and acidic soil produces blue-mauve flowers. Due to the flower formation, this species is referred to as a 'mophead' type, whereas the delicate-flowered *H. serrata* plants are known as 'lacecaps'. **HABITAT** Native to east Asia in Japan. Cultivated widely in cooler climes in India, such as the hill stations of Shimla and Northeast India. Thrives in partially shaded and full-sun locations.
USES Ornamental shrub grown for its attractive foliage and flowers. In China, leaves and roots are used medicinally as anti-malarial treatment. **ETYMOLOGY** Genus name is derived from the Greek *hydor*, water, and *aggeion*, vessel, in reference to the vessel-shaped flowers. The epithet *macrophylla* refers to the large leaves.

Bearded Iris ■ *Iris* x *germanica*

DESCRIPTION Slender plant that grows from rhizomes to 0.4m in height. Bright green leaves are strap shaped with sharp-pointed tips. Attractive flowers have an interesting floral

structure with three inner petals (standards) and three outer segments (falls). Generally, falls have markings to guide insect pollinators in, and the central fall is hairy. Flowers bloom in spring to early summer, and are often fragrant. They are borne in pairs on a tall stalk. **HABITAT** Bearded or German iris group is a hybrid of *I. germanica* with other European species. Thrives in temperate climate. **USES** Ornamental groundcover planted for its showy flowers. Some varieties thrive on water edges and can be used in bog or aquatic gardens. Orris root is derived from the rhizomes and is used as a herb in Middle Eastern cuisine, and also in perfume making. **ETYMOLOGY** Genus is named after the Greek rainbow goddess Iris. The epithet name refers to one of its parent species, which is from Germany.

Fringed Iris ■ *Iris japonica*

DESCRIPTION Evergreen perennial shrub that grow from rhizomes. Up to 0.6m in height. Leaves long and narrow, like straps, and arranged like a fan. Pale lavender flowers have fringed inner petals and yellow markings. They are borne solitarily on tall stalks and bloom in March–April. **HABITAT** Native to east Asia, in Japan and China. Cultivated in cooler climate, such as the valleys and hillsides of Kashmir, East India and Nepal. In its natural habitat, grows in forest margins and thrives in partial shade as an understorey plant. **USES** Suited for use as a groundcover in shaded areas under large trees. Cultivated for its attractive flowers. In its native region in China, rhizome is used in traditional medicine remedies to treat wounds. **ETYMOLOGY** Genus is named after the Greek rainbow goddess Iris, and epithet name refers to its origins in Japan.

Freesia ▪ *Freesia* hybrid

DESCRIPTION Herbaceous plants that grow upright, rising from corms to a height of 0.4m. Leaves lance shaped and erect, fanning out in one plane. Funnel-shaped flowers attractive and fragrant, arranged on a spike that is bent horizontally. They bloom in cold months, in December–March, and attract bees. Leaves dry out after flowering and corms remain dormant through summer. **HABITAT** Hybrid cultivars of parent species such as *F. alba*, *F. leichtlinii*, *F. corymbosa* and *F. refracta*, which are native to South Africa. Thrives in cool climate conditions. Widely cultivated in India as a winter seasonal plant in the plains, and during summer in the hills and mountains. **USES** Ornamental seasonal grown for its showy, fragrant flowers, which are long lasting and also used as cut flowers. Essential oil derived from flowers used in perfumes and cosmetics. **ETYMOLOGY** Genus is named after German physician Freidrich Freese.

Walking Iris ■ *Neomarica gracilis*

DESCRIPTION Evergreen perennial shrub with light green foliage. Can grow to 0.5m in height. Long, strap-like leaves arise from rhizomes and have a clump-forming habit.

Flowers similar to those of the Bearded Iris (see p. 92), with white petals that have blue markings. Scented flowers bloom in spring, in April–May, and are borne solitarily on a tall stalk. After flowering, the stalk with the seed pod bends to the ground to start a new shoot, hence the common name Walking Iris. **HABITAT** Native to Central America from Mexico to Brazil. Not common in India but cultivated in cooler climes of Northeast India and higher altitudes of Kathmandu. **USES** Ornamental shrub grown as a groundcover for its beautiful flowers. Also suited to growing in containers. **ETYMOLOGY** Genus name originates from *neo*, new, and *marica*, a genus from the Iridaceae family named after a river nymph. The epithet *gracilis* translates as slender or graceful.

Ironwood ■ *Mesua ferrea*

DESCRIPTION Slow-growing evergreen tree with a straight trunk and narrow crown. Can grow to 8m tall. Leaves narrow and lance shaped, simple and arranged in opposite pairs. Young leaves droopy and bright red or pink. Flowers white with yellow stamens in

centre. They appear in February–March and are borne solitarily or in pairs. Fragrant flowers attract bees and other insects for pollination. **HABITAT** Native to tropical Asia, including the Indian subcontinent. Grows in the Eastern Himalayas and Western Ghats. Thrives in warm, humid climate and tolerates partial shade. **USES** Grown in gardens for its attractive form, foliage and flowers. The hardwood was traditionally used for railway sleepers. Flowers have religious significance and are used as votive offerings. They are often planted near temples and known locally as *Nagkesar*. Most parts of the plant are used in traditional medicine systems such as Ayurveda, and are considered to be anti-inflammatory, antifungal and antibacterial. **ETYMOLOGY** Genus is named after Persian physician Yuhanna Ibn Masawayh, whose anglicized name was John Mesue. The epithet *ferrea* refers to the timber, which is supposedly hard as iron.

Cape Leadwort ■ *Plumbago auriculata*

DESCRIPTION Semi-deciduous shrub with a scandent habit, which can be trained as a climber and can grow to 2m in height. Bright green leaves oval in shape and arranged in opposite pairs. Flowers pale blue, salverform in shape, and with a long tube opening out flat to five petals. They bloom in summer to autumn and attract butterflies. **HABITAT** Native to South Africa. Cultivated widely in India and prefers a position in partial shade. **USES** Dense shrub with pretty flowers, often used as a groundcover in garden beds. Can also be trained as a climber along a fence. Suited to planting in butterfly gardens since it is a food source for butterflies. **ETYMOLOGY** Genus name is derived from the Latin *plumbum*, lead, named after a species in Europe that is used to treat lead poisoning. The epithet name is drawn from *auricula*, ear, due to the ear-shaped appendage at the base of the leaves.

White Leadwort ■ *Plumbago zeylanica*

DESCRIPTION Scrambling shrub with semi-deciduous foliage and a low, spreading habit. Grows to 1.5m in height. Leaves simple and oval in shape, arranged in opposite pairs. White flowers bloom in clusters and are salverform in shape, with a tubular end opening to five flat-faced petals. They flower in autumn, in October–November, and attract insect pollinators. **HABITAT** Native to India and south Asia. Grows wild in scrubland and grassland, and can thrive in semi-arid conditions. **USES** Hardy shrub with beautiful flowers. Not common in Indian gardens, but found in the wild or cultivated in biodiversity parks. Root is a source of plumbagin and known to be used in traditional medicines. **ETYMOLOGY** Genus name is derived from the Latin *plumbum*, lead, after a species in Europe that is used to treat lead poisoning. The epithet *zeylanica* refers to Ceylon or Sri Lanka, one of its native ranges.

Blood Wood Tree ■ *Haematoxylon campechiana*

DESCRIPTION Small tree with semi-deciduous foliage and spreading crown. Grows to 5–6m in height. Slow-growing plant with spiny branches. Leaves compound, evenly pinnate and arranged alternately. Yellow flowers minute and fragrant, densely packed along a spike. They bloom in dry months, in December–March. Nectar-rich flowers attract bees. **HABITAT** Native to Central America – Belize, Guatemala and Mexico. Not commonly cultivated in Indian gardens; thrives in hot and humid conditions of South India and West

Bengal. **USES** Grown in gardens for its fine foliage and flowers; commercially grown as log wood. Dye from heartwood is used to colour fabrics, and is also the source of haematoxylin, a staining agent used in the study of cells and tissues. Wood is also known to be used for medicinal purposes. **ETYMOLOGY** Genus name is derived from the Greek *haima*, blood, and *xylon*, wood, referring to the brown-red heartwood. The epithet *campechiana* refers to the Gulf of Campeche in Mexico, one of its places of origin.

Butterfly Pea ■ *Clitoria ternatea*

DESCRIPTION Slender evergreen climber with vigorous growth, which can grow to 5–6m in height. Leaves pinnately compound, and arranged alternately with oblong leaflets. Blue flowers have a pale-yellow centre, and are borne solitarily or in pairs. Some variants have pure white flowers. They bloom in summer and attract insect pollinators such as bees and butterflies. **HABITAT** Native to tropical and subtropical parts of Africa and Southeast

Asia. Widely cultivated in gardens throughout the Indian subcontinent, and popularly known as *Aparajita*. **USES** Ornamental climber trained over lattices and fences. Flowers often used as votive offerings. Edible, and can be used to brew tea or garnish salads. Used in traditional medicine systems such as Ayurveda. Roots, seeds and leaves used externally to treat skin ailments, and internally for stomach ailments. **ETYMOLOGY** Genus named after clitoris, referring to the shape of the flower, which resembles the human female genitalia. The epithet *ternatea* means cluster of three.

Blake's Coral Tree ■ *Erythrina blakei*

DESCRIPTION Small deciduous tree that often grows with a crooked trunk and dark, fissured bark, up to a height of 4–5m. Leaves compound and trifoliate with three triangular leaflets. Scarlet flowers borne on tall spikes in early spring, in February–April. Petals have a closed, curving form, with prominent stamens sticking out on the undersides.

HABITAT Known to be a hybrid cultivar, and popularly grown in gardens and parks throughout the Indian subcontinent. **USES** Cultivated as an ornamental tree in gardens and parks. **ETYMOLOGY** Genus name is derived from the Greek *erythros*, red, referring to the colour of the flowers. The epithet *blakei* was coined by its breeder, William Macarthur, after his gardener Edmund Blake.

India Coral Tree ■ *Erythrina variegata*

DESCRIPTION Handsome deciduous tree with a straight trunk and narrow crown. Up to 10m tall. Leaves heart shaped and compound, trifoliate, comprising three leaflets. Scarlet flowers attractive, and clustered on tall spikes. They bloom in early spring when the tree is nearly leafless, in February–April, and attract birds and insects. **HABITAT** Native to Southeast Asia, including the Indian subcontinent. Cultivated widely in the gardens of the region. **USES** Ornamental tree grown in parks and gardens for its attractive form and flowers. Leaves and bark used in traditional medicines in many parts of Asia. **ETYMOLOGY** Genus name is derived from the Greek *erythros*, red, referring to the colour of the flowers. The epithet *variegata* refers to the yellow-green variegated leaves of some variants.

Chinese Wisteria ■ *Wisteria chinensis*

DESCRIPTION Quick-growing deciduous climber with woody stems that twine around support structures. Grows to 10m in height. Leaves pinnately compound with oblong leaflets that are acutely tapered at the ends. Mauve-white flowers in long, drooping clusters bloom profusely in late spring, in April–May, and attract insect pollinators like bees and butterflies. *W. c.* var. 'Alba' has pure white flowers. All parts of the plant are toxic. **HABITAT** Native to China, and suited to subtropical and temperate climates. Cultivated at higher altitude, cold-climate regions of the Indian subcontinent, such as Kashmir, Uttarakhand, Sikkim and Bhutan. Adapts easily and is also becoming popular in the plains. **USES** Ornamental climber, mostly trained along a wall or on arbours. Considered invasive in some parts of the world, and its growth should be kept in check. **ETYMOLOGY** Genus is named after the eighteenth-century North American anatomist Casper Wistar. The epithet *chinensis* means from China.

Desert Cassia ■ *Senna polyphylla*

DESCRIPTION Large, semi-deciduous shrub, with wide, arching branches, which grows to 3m in height. Leaves pinnately compound, feather-like and arranged alternately with oval-shaped leaflets. Yellow flowers with five petals bloom in clusters nearly throughout the year, and attract bees and other insects. **HABITAT** Native to North and South America, Puerto Rico, Virgin Islands and Bahamas. Hardy plant that is widely cultivated in India. Prefers a location in full sun. **USES** Ornamental shrub grown for its attractive arching form and flowers. Can be planted close together as a hedge. **ETYMOLOGY** Genus name is derived from the Arabic *sana*, a vernacular name for a species with laxative leaves and pods. The epithet *polyphylla* refers to its numerous small leaflets.

Sturt's Desert Pea ■ *Swainsona formosa*

DESCRIPTION Annual herbaceous shrub with low-spreading, trailing habit. Grows to 0.3m in height. Known to be perennial in its natural habitat. Leaves pinnately compound and arranged alternately. Leaflets oblong and grey-green in colour. Attractive flowers unusual in shape, with crimson-red petals and black disc in centre. They are borne in clusters on an erect spike. **HABITAT** Native to arid and semi-arid regions of Australia. Not common but cultivated as a specimen winter seasonal in India. Thrives in well-drained, sandy soil in sunny location. **USES** Ornamental seasonal grown for its showy flowers, mostly as a potted plant. **ETYMOLOGY** Genus is named after English botanist Isaac Swainson, and the epithet *formosa* means handsome or beautiful. Common name is in honour of the explorer Charles Sturt, who recorded and described the plants in central Australia.

Java Cassia ■ *Cassia javanica*

DESCRIPTION Small deciduous tree with a spreading crown and drooping branches. Leaves compound and feather-like, with small, oval leaflets arranged pinnately. Pink white flowers appear in profuse clusters in April–May, and attract insect pollinators. **HABITAT** Native to Indonesia, Malaysia and Thailand. Widely cultivated in gardens throughout

the Indian subcontinent, thriving in mild climate. **USES** Ornamental shade tree planted as a specimen in gardens. Known to be used in traditional herbal medicines in its native region. Wood used for construction and making furniture. **ETYMOLOGY** Genus name is derived from the Greek *kassia*, a term used by the physicist Dioscorides for True Senna, which is aromatic or fragrant. The epithet *javanica* refers to its origins in Java, Indonesia.

Laburnum ■ *Cassia fistula*

DESCRIPTION Medium-sized deciduous tree with a spreading crown. Grows to 10m in height. Leaves compound with large, oblong leaflets arranged pinnately. Bright yellow flowers borne in large, drooping clusters covering the entire crown, and are fragrant. Flowers appear at onset of summer and last throughout the hot months through August–September. They attract birds and butterflies. Fruits are long, cylindrical pods that stay on the tree for a long time, making them a distinguishing feature of the tree. **HABITAT** Native to India, Pakistan, Sri Lanka and parts of Malaysia. Cultivated widely throughout the Indian subcontinent. **USES** Ornamental tree most often grown as an avenue tree.

Young leaves and flowers edible and can be cooked and eaten. All parts of the tree used medicinally and known to be antifungal, anti-inflammatory and laxative. Also used to treat wounds and skin ailments. **ETYMOLOGY** Genus name is derived from the Greek *kassia*, a term used by the physicist Dioscorides for True Senna, which is aromatic or fragrant. The epithet *fistula* is Latin for pipe, referring to the pipe-like cylindrical fruits.

Flame of the Forest ■ *Butea monosperma*

DESCRIPTION Slow-growing deciduous tree with irregular form and compact crown. Grows to 8–10m in height. Leaves compound and trifoliate, comprising three broad leaflets. Flowers flame-orange; flowering occurs profusely when tree is near leafless. Flowers have beak-like, curved petals. They bloom in spring, in March–April, and are rich in nectar, which attracts many kinds of birds. **HABITAT** Native to Asia, from the Indian subcontinent, east up to Vietnam and Indonesia. Widely distributed in the Indian subcontinent. Drought-tolerant, hardy tree that can adapt to difficult growing conditions. **USES** Cultivated in gardens and parks for its showy bursts of flowers. Dye from orange

flowers used to make colours for celebrating Holi – the festival of colours. Flowers, leaves, bark and seeds are medicinal and used in traditional medicine systems such as Ayurveda. Lac insects hosted by the tree produce shellac gum known as Bengal Kino. **ETYMOLOGY** Genus is named after John Stuart, Earl of Bute, who was a patron of botany. The epithet *monosperma* refers to the single seed contained in the seed pod.

Pride of Burma ▪ *Amherstia nobilis*

DESCRIPTION Medium-sized evergreen tree with a spreading crown and weepy branches. Grows to 8m in height. Lance-shaped leaves compound and arranged pinnately, and young leaves reddish and droopy. Beautiful drooping clusters of vermillion red flowers with spots of yellow on long stalk. Blooms in January–March, and known to be pollinated by insects. **HABITAT** Native to southern Myanmar up to Thailand. Cultivated in hot, moist climate regions in India, such as the state of West Bengal, and parts of South India, Myanmar and Bangladesh. **USES** Ornamental tree prized for its showy flowers and also as a shade tree. Some consider it to be the most beautiful flowering plant in the Indian subcontinent's gardens. **ETYMOLOGY** Genus named after Lady Sarah Amherst, an enthusiast botanist and plant collector. The epithet *nobilis* means noble or grand.

Powderpuff ▪ *Calliandra haematocephala*

DESCRIPTION Attractive evergreen shrub with a spreading crown. Grows to 3–4m in height. Feather-like leaves bipinnately compound, arranged alternately with oblong leaflets. Flowers a mass of closely packed red stamens resembling powder puffs. Variants with pink and white flowers also common. Flowers bloom intermittently throughout the

year, and attract sunbirds and butterflies. **HABITAT** Native to Bolivia in South America. Hardy plant cultivated in gardens throughout the Indian subcontinent. Prefers a location in full sun. **USES** Ornamental shrub grown for its showy flowers. **ETYMOLOGY** Genus name is derived from the Greek *kallos*, beautiful, and *aner*, male, referring to the scarlet stamens. The epithet *haematocephala* also refers to the blood-red flowers.

Poinciana ▪ *Caesalpinia pulcherrima*

DESCRIPTION Large, semi-deciduous shrub with a rounded crown of feathery foliage. Grows to 3m in height. Leaves bipinnately compound with small oval leaflets, and arranged in opposite pairs. Showy orange flowers tinged with red and yellow appear in tall clusters above crown. *C. pulcherrima* var. *flava* has purely yellow blossoms. Flowers bloom intermittently throughout the year, and attract bees and other insects. **HABITAT** Native

to south-east Mexico up to Central America. Hardy plant cultivated widely in India. Prefers a location in full sun. **USES** Ornamental, hardy shrub grown along streets, and in parks and gardens. Known to be a food plant for butterflies and suitable for butterfly gardens. **ETYMOLOGY** Genus is named after Italian botanist and physicist Andrea Caesalpinio. The epithet is from the Latin *pulcher*, which means beauty.

Purple Orchid Tree ■ *Bauhinia purpurea*

DESCRIPTION Medium-sized, semi-deciduous tree with irregular form. Grows to 8m in height. Simple leaves broad and rounded, with a deep cleft that divides them into two segments. Purple flowers have a hint of pink and white, and are similar in shape to orchids. They flower in autumn, in September–November, and attract bees. Similar species *B. variegata* has larger flowers with broad, overlapping petals in shades of white and pink. **HABITAT** Native to the Indian subcontinent up to south China, occurring in

the foothills of the Himalayas. Widely cultivated in India and does well in full sun and partial shade. **USES** Ornamental tree grown for its attractive flowers and as a shade tree. Flowers edible and can be cooked or pickled. Roots and flowers used in traditional medicine, and known to be carminative and laxative. **ETYMOLOGY** Genus named by Carl Linnaeus, honouring Swiss botanist twins Jean and Gaspard Bauhin. The epithet *purpurea* refers to the purple flowers.

Yellow Orchid Tree ■ *Bauhinia tomentosa*

DESCRIPTION Medium-sized, semi-deciduous tree with irregular form. Grows to 8m in height. Simple leaves broad and rounded, with a deep cleft that divides them into two segments. Purple flowers have a hint of pink and white, and are similar in shape to orchids. They flower in autumn, in September–November, and attract bees. Similar species *B. variegata* has larger flowers with broad, overlapping petals in shades of white and pink. **HABITAT** Native to the Indian subcontinent up to south China, occurring

in the foothills of the Himalayas. Widely cultivated in India and does well in full sun and partial shade. **USES** Ornamental tree grown for its attractive flowers and as a shade tree. Flowers edible and can be cooked or pickled. Roots and flowers used in traditional medicine, and known to be carminative and laxative. **ETYMOLOGY** Genus named by Carl Linnaeus, honouring Swiss botanist twins Jean and Gaspard Bauhin. The epithet *tomentosa* means hairy, referring to the velvety pods.

White Orchid Tree ▪ *Bauhinia acuminata*

DESCRIPTION Evergreen shrub with bushy form. Can grow to 2–3m tall. Leaves simple and arranged alternately. They are broad with a deep cleft in the middle, dividing them into two segments that have acutely pointed tips. Pretty white flowers star shaped, comprising five petals. They bloom from March throughout the warm summer months, and attracts bee and butterflies.

HABITAT Native to east and Southeast Asia, including the Indian subcontinent, and widely cultivated in gardens throughout India. Thrives in full sun and also does well in partially shaded locations. **USES** Ornamental shrub grown for its snow-white flowers. **ETYMOLOGY** Genus named by Carl Linnaeus, honouring Swiss botanist twins Jean and Gaspard Bauhin. The epithet *acuminata* refers to the sharp-pointed leaves.

Lupin ▪ *Lupinus polyphyllus*

DESCRIPTION Slender, bushy perennial shrub that can grow to 1m in height. Many garden hybrids, such as *L. nanus*, are dwarf-sized annuals. Leaves palmately compound, comprising 8–10 leaflets. Blue, mauve, pink or white flowers borne on tall spikes and resemble pea flowers. They bloom in spring and attract bees and butterflies. **HABITAT** Native to North America, and suited to subtropical and temperate climates. Grown as a perennial in regions with colder climates, such as Nepal and Kashmir. In plains, grown as a winter annual. **USES** Ornamental shrub that makes an attractive filler for garden beds. Fixes soil nitrogen by means of bacteria that dwell in the root system. **ETYMOLOGY** Genus name is derived from the Latin *lupus*, to wolf down, mistakenly referring to the belief that the plant depleted soil nutrients. The epithet *polyphyllus* means many leaves.

Rose of Venezuela ■ *Brownea coccinea*

DESCRIPTION Small evergreen tree with a straight trunk and dense, spreading crown. Grows to 8m in height. Branches pendant with pinnately compound leaves that are arranged alternately. Leaflets lance shaped, with young ones appearing in drooping, red-pink clusters. Bright red flowers tubular with yellow stamens projecting in centre. They appear in spring, in February–March, in round clusters borne drooping from tree trunks

and mature branches. **HABITAT** Native to South America from Venezuela to Ecuador. Cultivated in hot and humid regions of the Indian subcontinent such as South India, West Bengal and the Western Ghats. **USES** Ornamental shade and flowering tree grown in parks and gardens for its beautiful blossoms. **ETYMOLOGY** Genus is named after Irish botanist and physician Dr Patrick Browne. The epithet *coccinea* refers to the red flowers.

Sweet Pea ■ *Lathyrus odoratus*

DESCRIPTION Herbaceous annual vine that climbs with the help of curling tendrils and can grow to 2m in height. Compound leaves comprise two leaflets and are arranged alternately. Sweet-smelling flowers pea shaped and appear in clusters. They bloom in spring and occur in shades of white, pink, blue and purple. Fruit similar to a pea pod; it is known to be toxic. **HABITAT** Native to southern Europe, in Italy and Sicily. Widely cultivated in India as a winter seasonal plant. **USES** Popular winter seasonal grown for

its sweet-scented flowers, and trained along fences and borders. One of the earliest plants to be commercially grown for cut flowers. Essential oil from flowers used in perfumes and cosmetics. Root system is characteristic of all the pea family, and helps to fix nitrogen levels in soil. **ETYMOLOGY** Genus name is derived from the Greek *lathyros* for pea or pulse, and the epithet name *odoratus* refers to the fragrant flowers.

Sita Asok ▪ *Saraca asoca*

DESCRIPTION Small evergreen tree with straight trunk and dense, rounded crown. Can grow to 6–8m in height. Leaves pinnately compound and arranged alternately. Lance-shaped leaflets pink-red when young and drooping. Attractive flowers bloom profusely in February–April, and appear in large, round clusters on trunk or mature branches. Tubular flowers orange with tints of yellow when they open, turning red as they mature. Their fragrance is intense in the evening, and attracts bees and butterflies. **HABITAT** Native to tropical parts of the Indian subcontinent, and found in South India, the Western Ghats, Orissa, Assam and foothills of the Himalayas. Thrives in hot, humid climate in full sun or partial shade. **USES** Ornamental shade and flowering tree, planted in parks and along avenues. Considered auspicious by Hindus and Buddhists, and flowers are used as votive offerings. Stem, bark, flowers and seeds used in traditional medicine systems such as Ayurveda and homeopathy. **ETYMOLOGY** Origin of the genus name uncertain, and epithet name is derived from Indian vernacular name *Asoka*.

Yellow Saraca ▪ *Saraca thaipingensis*

DESCRIPTION Small evergreen tree with a dense rounded crown. Grows to 7–8m in height. Leaves pinnately compound, arranged alternately, with lance-shaped leaflets. Young leaves droopy, and pink-red in colour. Flowers cauliflorous, borne on tree trunks or on mature branches, with numerous florets appearing in densely packed clusters. They bear a faint fragrance and are rich in nectar, attracting bees, butterflies and sunbirds. They bloom in September–December. **HABITAT** Native to east Asia – Myanmar, Indonesia, Malaysia and Thailand. Cultivated in hot, humid regions in the Indian subcontinent, favouring a spot in partial shade. **USES** Ornamental shade tree grown as a specimen in gardens. Host plant for butterflies. **ETYMOLOGY** Origin of the genus name uncertain. The epithet *thaipingensis* refers to one of its origins in Taiping, Malaysia.

Himalayan Tulip ■ *Tulipa clusiana*

DESCRIPTION Annual herbaceous plant growing to 0.3m tall. Grows from a bulb, with leaves that are long, lance shaped and rise upright. Star-shaped flowers white with pink markings. They bloom in spring, in March–April, and are borne solitarily on tall stalks. Most garden tulips are hybrids and occur in various shapes and colours. Bicolour tulips are referred to as 'broken', and others of a single colour are called 'self coloured'. **HABITAT** *T. clusiana* is an Asian species of tulip native to Iran, Iraq, Pakistan and western Himalayas in India. It grows wild in the fields and slopes of Kashmir. **USES** Prized for its highly attractive blooms, and grown as a garden ornamental. The Tulip Garden of Srinagar in Kashmir is spread over acres of land and features more than 60 varieties. Flowers long lasting and make popular cut flowers.

Tulip hybrid varieties

Kaiser's Crown ■ *Frittilaria imperialis*

DESCRIPTION Attractive perennial shrub with an upright form, growing from a bulb to 1m tall. Lance-shaped leaves arranged in whorls along stem. Bell-shaped flowers orange with dark brown markings on insides. Clusters of drooping flowers borne on a tall stalk and topped by a crown of leaf-like bracts. They bloom in spring, in April–May, and attract bees for pollination. Flowers and bulbs carry an odour that is known to repel rodents. **HABITAT** Native to southern Turkey and Iran, up to north-west Himalayas. Grows wild in Kashmir and is cultivated in its gardens. **USES** Grown as a garden ornamental for its attractive form and flowers. **ETYMOLOGY** Genus name derives from the Latin *fritillus*, dice box, in reference to a chequered pattern that can be seen in the petals of the species *F. meleagris*. The epithet *imperialis* means of the emperor, in description of the crown-like bracts at the top of the flower cluster.

Crape Myrtle ■ *Lagerstroemia indica*

DESCRIPTION Small deciduous tree with a compact crown. Grows to 3–4m in height. Leaves small and oval shaped, and turn crimson in winter. Flowers profuse and appear from late spring to early winter, with colours ranging from white and pink, to mauve. Petals

crinkled, with crepe-like texture, hence the plant's common name. **HABITAT** Native to tropical and temperate parts of Asia, including India, and thrives in hot, humid climates. **USES** Ornamental tree grown in parks and gardens for its attractive blossoms. Flowers, stem, bark and leaves have medicinal properties, and are known to be astringent, stimulant and purgative. **ETYMOLOGY** Genus is named after Magnus von Lagerstrom of Goteborg, a Swedish naturalist who acquired many plants from India and China. The epithet refers to India, one of the plant's places of origin.

Queen's Crape Myrtle ▪ *Lagerstroemia speciosa*

DESCRIPTION Medium-sized, semi-deciduous tree with a rounded crown. Grows to 10–12m in height. Large, oblong leaves dark green and have prominent veins. Pink-purple flowers with crepe-like texture appear in large, showy clusters and fade to white with time. They bloom twice a year, in April–May and July–August, and attract insects such as bees for pollination. **HABITAT** Native to the Indian subcontinent, China, Malaysia and the Philippines. **USES**
Ornamental tree planted in parks and gardens for its beautiful flowers and shade-giving canopy. Often planted as an avenue tree. Root, bark, leaf and seed are used in medicines as an astringent and purgative. **ETYMOLOGY** Genus is named after Magnus von Lagerstrom of Goteborg, a Swedish naturalist who acquired many plants from India and China. The epithet *speciosa* refers to the showy flowers.

Lotus ▪ *Nelumbo nucifera*

DESCRIPTION Perennial aquatic plant that grows in water with its rhizomes rooted in mud. Young leaves float on the water and develop strong stems that rise above the water's surface. Mature leaves round with wavy edges, and rise to 1m in height. They have a water-repellent surface that does not allow moisture or dust to settle on it. Flowers rise high on sturdy stalks and open to a cup shape, with a central upturned cone surrounded by yellow stamens. They

are predominantly pink, or sometimes white, and are scented, attracting bees, flies and beetles. They bloom in summer, profusely during monsoons, in July–September. **HABITAT** Native to east Asia, including the Indian subcontinent. Widely distributed in the subcontinent from Kashmir in North India, Northeast and South India, as well as in Sri Lanka, Bangladesh and Pakistan. **USES** Ornamental plant grown in water gardens or large pots for its showy leaves and flowers. Flowers have religious significance in Hindu and Buddhist faiths and are used as votive offerings. They feature in religious iconography as a symbol of purity. Rhizomes edible, and eaten raw in salads or cooked. All parts of the plant are medicinal and used in traditional medicine systems such as Ayurveda. **ETYMOLOGY** Genus name is derived from its Sinhalese name. The epithet *nucifera* means nut producing.

Golden Champa ▪ *Magnolia champaca*

DESCRIPTION Medium-sized evergreen tree with a narrow crown. Grows to 6–8m in height. Elliptical leaves have sharp-pointed ends and are arranged alternately. Golden-yellow flowers known for their fragrance and borne solitarily. A popular garden hybrid, M. *alba*, bears creamy-white flowers. Blooms in summer, in June–July, and attracts birds and known to be pollinated by beetles. **HABITAT** Native to south Asia and southern China. Found in tropical and

subtropical moist forests of the Indian subcontinent from the Himalayas, Northeast India and up to Kerala and Sri Lanka. **USES** Ornamental tree grown for its fragrant flowers, used as votive offerings and often planted near temples. Flowers also worn by women to adorn the hair. Essential oil from flowers used in cosmetics and perfume making. Pale-coloured timber used in furniture making. Decoction of flowers used to treat coughs and colds. **ETYMOLOGY** Genus is named after French botanist Pierre Magnol, and the epithet *champaca* is derived from the Sanskrit name of the plant.

Lily Magnolia ■ *Magnolia liliiflora*

DESCRIPTION Deciduous shrub or small tree with multiple branches. Grows to 3–4m in height. Bright green leaves oblong and arranged alternately. Lily-like flowers are in shades of pink and are fragrant. Blooms profusely in late spring, in April–May. Known to be pollinated by beetles. **HABITAT** Native to south-west China, it is cultivated at higher altitudes and in the cooler climate of the Indian subcontinent, for example in Kashmir. In the plains, prefers a location in partial shade. **USES** Ornamental, bushy shrub grown for its showy blossoms. Buds and flowers used medicinally and known to be analgesic, sedative and a tonic. **ETYMOLOGY** Genus is named after French botanist Pierre Magnol, and the epithet *liliiflora* refers to the flowers, which resemble lilies.

Southern Magnolia ■ *Magnolia grandiflora*

DESCRIPTION Large evergreen tree with a dense crown. Grows to 10m in height. Oblong leaves glossy green with a leather-like texture, and arranged in groups of three along stem. Fragrant, creamy-white flowers bloom in summer, in June–September, and are known to be pollinated by beetles. One of the earliest known flowering plants, with an ancient flower structure. Fruit is followed by cone-like pod with bright red seeds. **HABITAT** Native to southeastern United States, and popularly cultivated in Indian gardens. Hardy tree that thrives in cooler climes, yet tolerates the hot, dry climate of North India, although the growth is stunted in such conditions. **USES** Ornamental tree grown for its attractive foliage and flowers. Essential oil from flowers used in perfume making. **ETYMOLOGY** Genus is named after French botanist Pierre Magnol, and the epithet, *grandiflora*, refers to the large, showy flowers.

Floss Silk Tree ■ *Ceiba speciosa*

DESCRIPTION Deciduous tree with spreading crown. Grows to 8m in height. Trunks of young trees green in colour and peppered with sharp spines. Leaves palmately compound, comprising five leaflets. Pink-white flowers showy and borne in loose clusters. The tree comes alive when in bloom, attracting sunbirds and bees in October–December. It marks

the onset of winter in the North Indian plains. **HABITAT** Native to South America, from Brazil, to Peru and Argentina. Widely cultivated in India as an ornamental in urban areas. Drought tolerant and requires a position in full sun. **USES** Popular ornamental in Indian gardens and parks; also planted as an avenue tree. Silky floss from its pods used as stuffing for pillows and cushions. **ETYMOLOGY** Genus name is derived from a vernacular South American name for Silk Cotton Tree, and the epithet *speciosa* means showy and refers to its splendid flowers.

Hibiscus ■ *Hibiscus rosa sinensis*

DESCRIPTION Medium-sized, bushy shrub, growing upright to 2m tall. Evergreen leaves are attractive, and are simple, dark green and glossy with serrated edges. Funnel-shaped flowers predominantly red, sometimes pink or white. They flower intermittently throughout the year, and attract sunbirds and other insect pollinators. **HABITAT** Native to India and tropical Asia, the plant's exact origins have not been determined. Cultivated in gardens throughout India, and requires a position in full sun. **USES** Popular garden ornamental, grown for its showy flowers that are also offered in prayer rituals and ceremonies. Closely planted and pruned, it makes an attractive hedge. Dried flowers used to make hibiscus tea, which is known to be a refreshing antioxidant drink. Used in traditional medicines of Ayurveda, it is astringent, cooling and an aphrodisiac. Dye made from petals is used in cosmetics for hair treatments. **ETYMOLOGY** Genus is named after the Greek *hibiskos* in reference to the true marsh mallow plant. The epithet *rosa sinensis* means rose of China.

Kenaf ■ *Hibiscus cannabinus*

DESCRIPTION Herbaceous shrub growing upright, with stems covered in prickly spines; up to 2m tall. Leaves deeply lobed into five segments. Flowers mauve, pale yellow or white with dark maroon centres. Funnel-shaped flowers borne solitarily and pollinated by bees.

HABITAT Origin of plant is presumed to be south Asia and also tropical Africa. Cultivated in Northeast India and locally known as *Sougri*. **USES** Grown popularly in kitchen gardens as a food crop in Northeast India; leaves cooked and eaten and are a great source of nutrition. Also grown as garden ornamental for the attractive flowers. Fibre produced from the plant has a jute-like quality. **ETYMOLOGY** Genus name is derived from the Greek *hibiskos* in reference to the true marsh mallow plant. The epithet *cannabinus* refers to the Cannabis-like leaves.

Kanak Champa ■ *Pterospermum acerifolium*

DESCRIPTION Large deciduous tree with a dense crown and straight trunk. Grows to 30m in height. Leaves simple, broad and rounded with wavy margins. They are silver-grey underneath and covered in fine hairs. Creamy-white flowers large and fragrant, and bloom in spring, in March–April. Buds open at night and attract bats for pollination. Seeds contained in a woody pod that opens into three segments to release them. **HABITAT** Native to the Indian subcontinent up to south China. Commonly found growing in

Himalayan foothills and parts of Central and East India. **USES** Tall, stately tree planted in parks and along avenues. Quick growing with dense foliage, and suited for planting as a screen. Flowers, bark and leaves used in traditional medicines. Fallen flowers are collected and kept in linen cupboards, due to their lingering fragrance and ability to repel insects. Leaves are sewn together to make plates. **ETYMOLOGY** Genus name is derived from the Greek *pteron*, wings, and *sperma*, seeds, referring to the winged seeds that are designed to be dispersed by the wind. The epithet *acerifolium* refers to the acer- or maple-like leaves.

Pink Ball ■ *Dombeya wallichii*

DESCRIPTION Large shrub with dense, rounded crown. Grows to 3m tall. Simple leaves are heart shaped with serrated edges, and arranged alternately. Pink-white flowers fragrant, and borne in rounded, drooping clusters. They bloom in March–May, and attract birds, bees and butterflies. **HABITAT** Native to Madagascar and popular in Indian gardens. Plant of tropical and subtropical climates, and thrives in full sun or partial shade. **USES** Ornamental shrub grown for its profuse, showy blossoms. Can be closely planted like a hedge to create a screen. **ETYMOLOGY** Genus is named after French botanist Joseph Dombey, and the epithet is named after Danish botanist Nathaniel Wallich.

Silk Cotton Tree ■ *Bombax ceiba*

DESCRIPTION Large deciduous tree with a straight trunk supported by large buttresses. Grows to 30m in height. Leaves palmately compound, arranged in groups of five leaflets. Flowers large and conspicuous, cup shaped, and borne in clusters in January–March. Yellow, red or orange petals thick and waxy; they open at midnight and last for a day. When in bloom, the tree is alive with birds, and bees and other insects. Also known to be pollinated by bats. **HABITAT** Native to tropical and subtropical Asia, including India, tropical Africa and northern Australia. Cultivated widely in the Indian subcontinent, it is drought tolerant and requires a location in full sun. **USES** Ornamental tree grown in parks and gardens; also used as an avenue tree. Petals and calyx of the flower can be cooked and eaten. Floss from the seed pods used as stuffing for cushions and pillows. All parts of the plant are medicinal. **ETYMOLOGY** Genus name is derived from the Greek *bombyx*, cotton, in reference to the fibre attached to the seeds. The epithet name *ceiba* refers to a group of trees such as *Ceiba pentandra*, which is also a source of silk cotton.

Princess Flower ■ *Tibouchina urvilleana*

DESCRIPTION Evergreen shrub with attractive foliage and upright stems, which grow to 2m in height. Lance-shaped leaves simple and arranged in opposite pairs. They are covered in fine hairs and have a velvet-like texture. Attractive purple flowers are saucer shaped, comprising five rounded petals. They bloom in July–August and are short lived,

lasting for a day. **HABITAT** Native to South America in Brazil, and thrives in hot and humid climate in South India, West Bengal and Northeast India. **USES** Ornamental shrub cultivated for its attractive foliage and flowers. **ETYMOLOGY** Genus name is derived from vernacular Guiana term for the plant. The epithet *urvilleana* is named after French explorer and botanist Jules Sebastien Cesar Dumont d'Urville.

Bleeding Glory Bower ■ *Clerodendrum thomsoniae*

DESCRIPTION Evergreen climber with dense, bushy foliage. Grows to 3m in height. Oblong leaves have tapered ends and prominent veins. They are simple and arranged in opposite pairs. Attractive flowers have an usual form, comprising four white calyces, four red petals and prominent white stamens. They bloom in the rainy season and attract insect

pollinators. **HABITAT** Native to West Africa, and cultivated widely in India, thriving in moist soil conditions. **USES** Grown as a garden ornamental for its jewel-like flowers. **ETYMOLOGY** Genus name is derived from the Greek *kleros*, chance, and *dendron*, tree (tree of chance). The epithet *thomsoniae* honours the wife of the Rev. William Cooper Thomson Jr., who was a missionary in Africa and first discovered and documented the plant.

Flaming Glory Bower

■ *Clerodendrum splendens*

DESCRIPTION Evergreen climber with a twining habit and dense foliage. Grows to 4–5m in height. Leaves glossy green, simple and arranged in opposite pairs, and have wavy margins. Scarlet flowers bloom abundantly in cold months, in October–March, in dense clusters. They are salverform in shape, with five petals that open flat and a long, tubular end, and attract sunbirds, bees and butterflies. **HABITAT** Native to western tropical Africa, and widely cultivated in gardens throughout the Indian subcontinent. A position in full sun encourages profuse flowering. **USES** Ornamental climber trained over fences and trellises, and appreciated for both its foliage and flowers. **ETYMOLOGY** Genus name is derived from the Greek *kleros*, chance, and *dendron*, tree (tree of chance). The epithet *splendens* means splendid or striking, referring to its spectacular blossoms.

Hill Glory Bower ■ *Clerodendrum infortunatum*

DESCRIPTION Woody evergreen shrub with dense foliage. Can grow to 2–3m tall. Heart-shaped leaves have serrated edges and are simple, arranged in opposite pairs. Flowers showy, appearing in large, tapering clusters on a spike. They bloom in April–June. Petals white with dark pink centres and long stamens. Seeds are prominent black beads set in red calyces. **HABITAT** Native to east Asia, in India, Sri Lanka, Bangladesh, Thailand and Malaysia.

Thrives in hot, humid climate in a position in full sun. **USES** Grown as an ornamental for its attractive flowers. Important medicinal plant, locally known as *Bhant*. All parts of the plant are used in traditional medicine systems such as Ayurveda, Siddha and Unani. Used to treat wounds and stomach ailments. **ETYMOLOGY** Genus name is derived from the Greek *kleros* chance, and *dendron*, tree (tree of chance). The epithet *infortunatum* means unfortunate, which may refer to its toxic qualities.

Mandarin's Hat
■ *Holmskioldia sanguinea*

DESCRIPTION Scrambling shrub with a scandent habit. Can be trained as a climber and can grow to 3m in height. Its oblong leaves are simple and arranged in opposite pairs. Flowers are of an interesting shape, often described as a Chinese or Mandarin hat. They bloom in cold months, in November–February, and are predominantly red in colour. However, orange and yellow variants commonly occur. **HABITAT** Native to the Himalayan range in the Indian subcontinent up to Myanmar. Plant of subtropical climate and prefers a location in full sun. **USES** Grown as a specimen plant for its curiously shaped flowers, and can also be maintained as a hedge. **ETYMOLOGY** Genus is named after Danish botanist Johan Theodor Holmskjold, and the epithet *sanguinea* means blood, referring to the dominant colour of the species.

Parrot's Beak ■ *Gmelina philippensis*

DESCRIPTION Woody evergreen shrub with a scandent habit. Grows to 3m in height. Simple leaves irregular in shape and arranged in opposite pairs. Flowers unusual, appearing at the end of a long, drooping spike of overlapping bracts. They bloom in summer months, in May–October, and are pollinated by insects. **HABITAT** Native to east Asia, Myanmar, Thailand, Malaysia, Singapore and the Philippines. Naturally occurring in forests thickets, it is an understorey plant. Cultivated widely in Indian gardens, and thrives in partial shade as well as in full sun. **USES** Popular in the Indian subcontinent's gardens, often as a climber over courtyard walls and trellises. **ETYMOLOGY** Genus is named after German naturalist Johann G. Gmelin. The epithet refers to the Philippines, which is one of its native regions.

Tulsi ■ *Ocimum tenuiflorum*

DESCRIPTION one of the most important plants of India, this is a semi-deciduous shrub of medium size with a rounded form, growing to 0.5–1m in height. Oval-shaped leaves highly aromatic and arranged in opposite pairs. Leaves and flowers of another cultivar are tinged purple; locally known as *Krishna Tulsi*. Flowers minute, greenish-white and borne on slender spikes. They attract insect pollinators such as bees. **HABITAT** Native to tropical and subtropical parts of the Indian subcontinent. Widely cultivated in India. **USES** Grown as a venerated plant of religious and medicinal significance. Considered sacred by Hindus and grown in home courtyards and temples. Ceremonies and rituals are conducted around it. Used in Ayurvedic medicines, it is a purifier, insecticidal and antibacterial. Rejuvenating herbal tea is made from its leaves. Stems are cut and made into rosaries. **ETYMOLOGY** Genus name is derived from the Greek *okimon*, referring to an aromatic herb; the epithet *tenuiflorum* refers to its slender floral inflorescence.

Bottlebrush ■ *Callistemom viminalis*

DESCRIPTION Medium-sized, semi-deciduous tree with a straight, fissured trunk. Up to 8m tall. Narrow, lance-shaped leaves simple and arranged alternately. Flowers showy and look like a brush used for cleaning bottles. They comprise pale, inconspicuous petals and bright red bristles of stamens, and are densely packed along a spike. They bloom in spring, in February–May. **HABITAT** Native to east coast of Australia. Widely cultivated

in gardens throughout the Indian subcontinent. Thrives in moist soil conditions, along waterbodies and streams. Prefers a full-sun location, and also tolerates partial shade. **USES** Cultivated in parks and gardens for its showy flowers. Essential oil is derived from its leaves. **ETYMOLOGY** Genus is named after the Greek *kallos*, beautiful, and *stemon*, stamens, referring to the prominent stamens. The epithet *viminalis* is derived from *vimen*, which means long, flexible shoots.

Nasturtium ■ *Tropaeolum majus*

DESCRIPTION Annual herbaceous plant with low-growing, scandent habit, which can be trained on a support. Rounded leaves simple, with each leaf at the end of a long, trailing stem. Yellow to orange flowers, borne solitarily in cold winter months, attract bees and other insects. **HABITAT** Native to Peru in South America, and cultivated widely in Indian

gardens as a winter seasonal. **USES** Popular ornamental seasonal in gardens and parks. Leaves and flowers edible, vitamin rich and used in salads. Often used in herbal medicines, and known to be antibiotic, antifungal and antiseptic. Good plant for kitchen gardens as it draws away pests and aphids from crops. **ETYMOLOGY** Genus was named by the Swedish botanist Carl Linnaeus after the Greek word for trophy, *tropaion*.

Angel's Trumpet ▪ *Brugmansia suaveolens*

DESCRIPTION Woody shrub with evergreen crown, growing to 2–3m in height. Simple leaves elliptical with tapered ends, and arranged alternately. Showy flowers trumpet shaped, borne singularly and drooping. Fragrance is intense in the evenings and is known to attract bats at night. Hybrid variants such as *B. insignis* have pink, yellow and orange flowers. Flowers bloom sporadically through warm summer months. Every part of the plant is poisonous. **HABITAT** Native to south-eastern Brazil. Popularly cultivated in cooler climes of the Indian subcontinent, for example in Northeast India and Nepal, where it grows along roadsides. **USES** Ornamental shrub grown for its showy flowers. Makes a good specimen plant in a garden. **ETYMOLOGY** Genus is named after Dutch naturalist Sebald Justin Brugmans; the epithet *suaveolens* means sweet smelling.

Chalice Vine ■ *Solandra maxima*

DESCRIPTION Large evergreen vine with woody growth habit. Grows to 10m in length. Oblong leaves simple and arranged alternately. Cup-shaped flowers large, showy and yellow at first, deepening to gold with time, and marked with deep purple veins. They bloom in summer with a night fragrance, and are known to attract bat pollinators. **HABITAT** Native to tropical America – Mexico, Venezuela and Colombia. Not commonly cultivated in

India but seen in gardens at higher altitudes and cooler climes, such as those in Sikkim and Nepal. **USES** Ornamental vine grown as a specimen plant for its attractive flowers. Used in Mexico in traditional medicines and rituals, and is a strong hallucinogen attributed with magical properties. **ETYMOLOGY** Genus is named after Swedish naturalist Daniel Solander, and the epithet *maxima* means of great size, referring to the large flowers.

Lady of the Night ■ *Brunfelsia americana*

DESCRIPTION Evergreen shrub that can grow to a small tree up to 3m tall in an ideal environment. Oblong leaves simple and arranged alternately. White flowers fade to cream with time; they are salverform in shape with a tubular end opening flat out to five lobed petals. Borne solitarily, their scent is most intense at night to attract nocturnal insects such as moths for pollination. They bloom in spring in February–April. All parts of plants of the Solanaceae family are toxic. **HABITAT** Native to tropical America, from the Caribbean to Venezuela in South America. Cultivated in the Indian subcontinent's gardens, thriving in partial shade. **USES** Ornamental shrub used as a filler in garden beds; essentially grown for its fragrant flowers. **ETYMOLOGY** Genus is named after German monk and botanist Otto Brunfels. The epithet *americana* means of the Americas.

Yesterday Today Tomorrow ■ *Brunfelsia pauciflora*

DESCRIPTION Evergreen shrub with bushy growth. Can grow to 2m tall. Leaves simple, oblong and alternately arranged. Purple flowers fade to mauve, then white, covering the shrub in tricoloured blooms. They are salverform, with tubular ends opening out to five-lobed petals, and fragrant, blooming in spring and attracting butterflies. **HABITAT** Native to tropical America (Brazil). Widely cultivated in gardens of the Indian subcontinent, favouring a spot in partial shade. **USES** Ornamental shrub planted for its fragrant tricoloured flowers. Food plant for butterflies, suited to growing in butterfly gardens. **ETYMOLOGY** Genus is named after German monk and botanist Otto Brunfels. The epithet *pauciflora* means few or scant flowering.

Night Blooming Jasmine ■ *Cestrum nocturnum*

DESCRIPTION Bushy evergreen shrub with multiple drooping branches. Grows to 2–3m in height. Simple leaves elliptical and arranged alternately. Greenish-white flowers salverform

with a tubular end opening out to fives lobes, and appear in drooping clusters. Buds open at night and are highly scented, attracting nocturnal moths and other insects. A similar species, C. *diurnum*, has an upright form and floral spike, and as the name suggests, blooms during the day. All parts of the plant are toxic. **HABITAT** Native to Central America, in Panama, Mexico and Cuba. Widely cultivated in gardens throughout the Indian subcontinent. Thrives in sunny locations and also suited to partial shade. **USES** Popularly cultivated in gardens for its night-scented flowers, and in Hindi- and Urdu-speaking areas known as *Raat ki Rani*. Food plant for butterflies. Recommended for growing in butterfly gardens and moonlight gardens. **ETYMOLOGY** Genus name is from the Greek *kestron*, from which a similar plant named *Kestrum* is derived. The epithet name *nocturnum* refers to its night-blooming flowers.

Day Blooming Jasmine, Cestrum diurnum

Potato Tree ▪ *Solanum wrightii*

DESCRIPTION Small evergreen tree with a low, spreading crown. Grows to 4m in height. Large leaves unevenly lobed, arranged alternately and simple. They are covered with prickly thorns on the undersides along the midribs. Star-shaped flowers borne in clusters are purple in colour and fade to white with time. Blooms appear intermittently throughout the year. **HABITAT** Native to tropical America, from Brazil and Bolivia. Cultivated in Indian subcontinent gardens as an ornamental tree. Thrives in tropical climate with a location in full sun. **USES** Cultivated as a specimen plant in gardens for its unusual foliage and showy flowers. **ETYMOLOGY** Genus name is assumed to be derived from the Latin *solamen*, to give comfort or relief, in reference to the sedative effects of the nightshade genus. The epithet is named after American botanist Charles Wright.

Thorn Apple ■ *Datura metel*

DESCRIPTION Annual herbaceous shrub with a bushy, spreading crown. Grows to 1m in height. Leaves simple and lance shaped, and arranged alternately. Funnel-shaped flowers white, with variants in purple, and borne solitarily in upright stalks. Cultivar *D. metel* 'Flore Pleno' is double flowering. Buds open in the evenings accompanied by a fragrance, and attract nocturnal moths. Summer blooming, the flowers also attract butterflies and

other insects. All parts of the plant are toxic. **HABITAT** Known to have originated in east Asia (India and South China), its native range is also speculated to be in tropical America (Texas to Colombia). The exact origin remains uncertain. White-flowered variant grows wild along roadsides and wasteland in the Indian subcontinent. Purple-flowered variant is often seen in cultivation. **USES** Mainly cultivated as a medicinal plant, and used in traditional medicines of Ayurveda and Siddha systems. Known to be anaesthetic, hypnotic and a hallucinogen. Flowers and fruits associated with the worship of the Hindu god Shiva, and used in votive offerings.

Mickey Mouse Bush ■ *Ochna serrulata*

DESCRIPTION Erect shrub with an evergreen, compact crown. Grows to 2m in height. Simple leaves elliptical with pointed ends and finely serrated margins, and arranged alternately. Fragrant yellow flowers bloom in spring, in February–March. Flowers followed by attractive red-coloured seed cups, with fruits that are green and turn jet-black. These attract birds and are a distinguishing feature of the plant. The red seed cup looks like Mickey Mouse, the Disney character. **HABITAT** Native to the Indian subcontinent, and widely distributed in India. **USES** Popular ornamental shrub in Indian gardens for its attractive flowers and fruits. Bark and roots used medicinally. **ETYMOLOGY** Genus name is derived from the Greek *ochne*, wild pear, since the leaves share a likeness to it.

Common Lilac ■ *Syringa vulgaris*

DESCRIPTION Small deciduous tree with a narrow crown and compact form. Grows to

4–5m in height. Dark green, glossy leaves simple, oval or heart shaped, and arranged in opposite pairs. Sweet-smelling flowers purple-pink, small, numerous and packed densely on a spike. They bloom in spring in April–May, and attract bees and butterflies. **HABITAT** Native to south-east Europe. Not common in cultivation in the Indian subcontinent except in cold, high-altitude places such as Kashmir. Prefers a location in full sun. **USES** Garden ornamental grown for its showy fragrant flowers. Essential oil from flowers used in perfumes and aromatherapy. **ETYMOLOGY** Genus name is derived from the Greek *syrinks* or *syrinx*, which means a hollow tube or pipe, referring to hollow stems filled with pith. The epithet *vulgaris* is Latin for common.

Coral Jasmine ■ *Nyctanthes arbor tristis*

DESCRIPTION Small deciduous tree with a narrow crown, growing upright to 5–6m in height. Leaves simple, arranged in opposite pairs, irregular in shape and predominantly heart shaped. Their texture is rough and dry, like sandpaper. Small white flowers appear in delicate clusters in autumn, in August–November. They are salverform in shape with dark orange tubes at the ends. They are night blooming, with an intense fragrance, and drop to the ground, forming a floral carpet around the tree by morning. They attract nocturnal moths and bees. **HABITAT** Native to south Asia, including India, Pakistan and Nepal. Widely cultivated in gardens throughout the Indian subcontinent. **USES** Popular in Indian gardens, and planted for the fragrant blossoms, which are often used as votive offerings. Flowers used to brew tea. Dye from the orange tubes of the flowers produces a saffron colour. Essential oil is derived from the flowers. Used in Ayurvedic medicines, it is considered anti-inflammatory, antiviral and antifungal. **ETYMOLOGY** Genus name is derived from the Greek *nyx*, night, and *anthos*, flower, referring to the night-blooming flowers. The epithet *arbor tristis* means tree of sorrow.

Arabian Jasmine ▪ *Jasminum sambac*

DESCRIPTION Evergreen rambling shrub with a scandent habit, which can also be trained as a climber to grow to up to 3m in height. Simple leaves broadly oblong and arranged in opposite pairs. White flowers bloom in summer and are fragrant, and borne in clusters. There are many cultivars of single, semi-double and double-flowering varieties.
HABITAT Considered to be native to India, and widely cultivated in gardens throughout the Indian subcontinent. Commercially cultivated in South India for its flowers, with the best-quality flowers produced in the city of Madurai. **USES** Locally known as *Mogra*, it is

popularly grown in gardens for its fragrant blossoms. They are used as votive offerings, made into garlands and worn to adorn hair. Essential oil from flowers used in perfumes and cosmetics. Leaves, flowers and roots used in traditional medicine remedies.
ETYMOLOGY Genus name is derived from a vernacular Arabic name, *Yasmin*, which translates as the gift of God. The epithet *sambac* is drawn from the Arabic term *zanbaq* for jasmine flower oil.

Common Jasmine ■ *Jasminum officinalis*

DESCRIPTION Deciduous climber that grows vigorously to 4–5m in height. Leaves pinnately compound, comprising 5–7 leaflets that have acutely pointed ends. Pink buds open to white flowers in summer; they are star shaped with tubular ends. They are intensely fragrant and attract bees and butterflies. **HABITAT** Spread over a wide range from Turkey to China, including the Himalayan range in the Indian subcontinent. Thrives in subtropical climate and prefers a location in full sun. **USES** Popularly grown for its fragrant flowers, which are also known locally as *Chameli*. They are an essential part of the Indian way of life and used as votive offerings, made into garlands, and used in wedding ceremonies and to adorn hair on a daily basis. Essential oil from flowers used in perfumes and cosmetics. Flowers also used to make jasmine tea. Leaves, flowers and roots used in traditional medicine remedies. **ETYMOLOGY** Genus name is derived from a vernacular Arabic name, *Yasmin*, which translates as the gift of God. The epithet *officinalis* means sold in shops, referring to its medicinal uses.

Golden Bell ■ *Forsythia viridissima*

DESCRIPTION Deciduous shrub with upright growth and spreading crown, growing to 2–3m in height. Leaves tiny and simple, elliptic and arranged in opposite pairs. Yellow flowers bloom profusely in spring, in March–April, and are arranged densely along bare branches before new leaves appear. They attract insects for pollination. *F. suspensa*, with drooping branches, is also popularly cultivated. **HABITAT** Native to southern China, and suited for planting in temperate climate. Cultivated at higher altitudes in the Indian subcontinent, for example in Kashmir, Bhutan and Sikkim. Thrives in a sunny location. **USES** Showy garden ornamental planted for its dramatic blossoms. Can also be closely planted as a hedge. **ETYMOLOGY** Genus is named after William Forsyth, the Scottish horticulturist and one of the founders of the Royal Horticultural Society. The epithet *viridissima* refers to the green stems.

Fragrant Olive ▪ *Osmanthus fragrans*

DESCRIPTION Evergreen shrub that can grow to be a small tree 3–4m tall, in ideal conditions. Dark green leaves simple, elliptical in shape, with fine, serrated edges and arranged in opposite pairs. Small, creamy-white flowers highly fragrant and appear in clusters. They bloom in summer, in June–August, and are visited by bees and flies. **HABITAT** Native to Himalayas and temperate parts of China and Indo-China. Suitable for planting in partially shaded locations. **USES** Ornamental shrub grown for its fragrant flowers and glossy foliage, and especially valued in winter gardens. Flowers used to flavour tea in China. Essential oil from flowers used in perfumes. Flowers, stem and bark used in traditional medicines. **ETYMOLOGY** Genus name is derived from the Greek *osma*, fragrance, and *anthos*, flowers. The epithet *fragrans* further reinforces the fragrant nature of the flowers.

Primrose Jasmine ▪ *Jasminum mesnyi*

DESCRIPTION Sprawling evergreen shrub with arching stems. Grows to 2–3m in height. Leaves compound and arranged in opposite pairs. They are trifoliate, comprising three acutely pointed leaflets. Yellow flowers bloom in March–May and are borne solitarily. They are salverform in shape with petals opening out flat, and have a tubular end. **HABITAT** Native to southwestern China and Vietnam. Thrives in mild, subtropical climate and does well in full sun and also in partial shade. **USES** Cultivated in gardens for its profuse yellow blossoms. Can also be planted closely to make a hedge. **ETYMOLOGY** Genus name is derived from a vernacular Arabic name, *Yasmin*, which translates as the gift of God. The epithet *mesnyi* is named after General William Mesny, who served in China and collected plant specimens to send back to Britain.

Blue Vanda ■ *Vanda coerulea*

DESCRIPTION Epiphytic herbaceous orchid growing on tree trunks, with numerous, silvery, aerial roots dropping down. Leaves leathery and strap-like. Known for its delicate clusters of blue flowers on branching spikes; variants with pink or white flowers also occur. Shade of blue varies according to the position of the plant – in full sun it is clear blue, and in partial shade it is blue with striations. Blooms in autumn, in September–October.

HABITAT Native to Northeast India, and also found in Nilgiri Hills in South India, Yunnan in China, Myanmar and Thailand. **USES** Valued for is showy blue flowers, and used by horticulturist to produce other hybrid orchids. Leaves and stem used in herbal medicine remedies. **ETYMOLOGY** Genus name is known to be derived from Sanskrit name *Vandaka*. The epithet *coerulea* refers to the blue colour.

Golden Bow Dendrobium ■ *Dendrobium chrysanthum*

DESCRIPTION Herbaceous epiphyte that grows on trees and draws sustenance from them. Leaves thick and leathery, lance shaped and cling to tree trunks by means of aerial roots.

Flowers golden-yellow with dark brown spots at centre, borne in drooping clusters. Another species that is often confused with this one is *D. chrysotoxum*, also known as *Khogunmelei* in vernacular Manipuri, differentiated by the absence of a dark brown flower centre. Blooms in summer during May–June. **HABITAT** Native to foothills of Himalayas in the Indian subcontinent, Myanmar, Thailand, Laos and Vietnam, up to China. Commonly cultivated in gardens in Northeast India. **USES** Ornamental plant grown on tree trunks for its showy flower clusters, or as a decorative hanging on tree stumps. Used in traditional medicines in the Indian state of Meghalaya. **ETYMOLOGY** Genus name is derived from the Greek *dendron*, tree, and *bios*, life, referring to its epiphytic nature of growing on trees. The epithet *chrysanthum* refers to its golden, or *chryseus*, colour.

Passion Flower ▪ *Passiflora incarnata*

DESCRIPTION Evergreen, fast-growing climber that latches on to supports by means of tendrils. Can grow to 6m in height. Leaves deeply lobed, divided into three or five segments. They are simple and arranged alternately. Flowers unusually attractive, radial with five inner petals and outer sepals enclosing numerous thread-like, wavy coronas. They bloom in summer and are pollinated by birds and bees. Fruits of certain species, such

as *P. edulis*, are aromatic and edible. **HABITAT** Native to North America – the south-east region from Florida to Texas. Widely cultivated in gardens throughout the Indian subcontinent. **USES** Grown for its attractive flowers, and best trained on trellises and fences. Known to be used in herbal remedies and ritual ceremonies in its native region. **ETYMOLOGY** Genus name is derived from the Latin *passion* and *flos*, flower, and it was named by Spanish missionaries to the Americas. They compared the likeness of the flower parts to the crown of thorns representing the passion of Christ.

Tree Peony ▪ *Paeonia suffruticosa*

DESCRIPTION Perennial deciduous shrub with bushy form. Grows to 2m in height. Leaves bipinnately compound and arranged alternately, with deeply lobed leaflets. Large, showy flowers pink-white, with dark centre surrounding yellow stamens. They bloom in late spring, in April–May. Some variants are fragrant and attract insect pollinators. **HABITAT** Native to Kashmir, Bhutan, Tibet and parts of China. Plant of temperate climates and cultivated in higher altitude places like Sikkim and Kashmir in India. Thrives in a sunny location. **USES** Ornamental shrub grown for its foliage and flowers. In China, used in traditional medicine remedies, and a liquor is made from fermented petals, known as Moudan wine. **ETYMOLOGY** Genus is named after the Greek *paeon* or physician of the gods, due to the medicinal properties of peonies. The epithet *suffruticosa* means shrubby.

Common Poppy ■ *Papaver rhoeas*

DESCRIPTION Annual herbaceous plant that grows to 0.5m tall on slender stems that

need to be supported. Delicate, cup-shaped flowers are red with crepe paper-like textured petals. They are borne on long stalks covered in fine hairs, are occasionally pink or white, and have a dark patch at the centre. They bloom in spring, in February–April, and attract bees. **HABITAT** Native to Eastern Mediterranean, southern Europe, Africa and temperate Asia. Widely cultivated in India as a winter seasonal. **USES** Popular seasonal plant grown in fields or borders in plant beds for its showy flowers. Flowers preserve well when pressed and dried. They are used traditionally to relieve pain. Poppy seeds are edible and used in cooking and baking, and are locally known as *Khus*. **ETYMOLOGY** Genus name is from *pappa*, the Latin term for food or milk, referring to the white latex that oozes from the stems when they are cut. The epithet *rhoeas* refers to the red flowers.

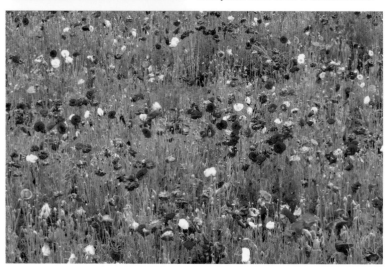

Poppy hybrid varieties

California Poppy ■ *Eschscholzi californica*

DESCRIPTION Annual herbaceous shrub that grows to 0.3m tall. Blue-green leaves feathery, finely divided and arranged alternately. Yellow-orange flowers cup shaped and borne solitarily on delicate stalks. They bloom in February–March in the plains during summer, and attract bees and other insects for pollination. **HABITAT** Native to North America, in California, Nevada and New Mexico, growing in open grassland in its native habitat. Cultivated as a seasonal plant in India, it thrives in well-drained, sandy soil in a sunny location. **USES** Ornamental flowering annual grown in borders or as a bedding plant. Hardy plant that can be used in creating wildflower gardens. In its native region, used medicinally as a sedative. **ETYMOLOGY** Genus is named in honour of Johann Friedrich Eschscholtz, an Estonian surgeon and naturalist who was part of an expedition to the Pacific Coast of North America. The epithet refers to its origin in California.

Cyclamen ■ *Cyclamen persicum*

DESCRIPTION Herbaceous perennial growing from tuberous roots to 0.3m tall. Patterned leaves simple and heart shaped, with fine, serrated edges. They drop after flowering and the plant remains dormant throughout summer to revive again in autumn. Flowers appear in late winter to spring, in December–March, in profuse blooms with upturned petals that have a twirl. Their colours include pale lavender, pink and sometimes white. **HABITAT** Native to North Africa – Algeria to east Mediterranean in Turkey. Thrives in relatively cool climate areas in the Indian subcontinent, such as the northeastern states, Sikkim, Kashmir and the Himalayas in Uttarakhand, and Himachal Pradesh. **USES** Popular as a houseplant grown in pots, and prized for its attractive flowers. **ETYMOLOGY** Genus name is derived from the Greek *kiklos*, circle, referring to the flat, round tuberous bulb. The epithet *persicum* means from Persia, where it was believed to have originated.

Primrose ■ *Primula vulgaris*

DESCRIPTION Low, compact perennial shrub that grows to 0.3m tall. Leaves large, glossy green with a crinkled texture, and arranged in rosettes. Flowers predominantly pale yellow with bright yellow centre. They bloom in winter to spring, in December–April, and attract insect pollinators such as bees, butterflies and moths. Hybrid versions such as Polyanthus varieties are now common garden variants and occur in different colours – pink, yellow, red and blue. **HABITAT** Native to Europe up to western Asia, and widely cultivated as a winter seasonal plant in the Indian subcontinent's gardens. **USES** Ornamental shrub grown as a winter seasonal for its bright blossoms. **ETYMOLOGY** Genus name is derived from *Primula veris* for daisy, meaning 'firstling of spring'. The epithet *vulgaris* means common.

Apple ■ *Malus domestica*

DESCRIPTION Small deciduous tree grown for its popular edible fruits. Cultivated trees grown on rootstock can grow to 3–5m in height. Oblong leaves have fine, serrated edges, and are simple and arranged alternately.

Blossoms appear in spring along with the new leaves, and have white petals with a tinge of pink. They attract many types of insect pollinator. **HABITAT** A hybrid of species, mainly M. *sylvestris* and M. *pumila*, it is cultivated in the Himalayas across Pakistan, Bhutan, India and Nepal, originating from the mountains of Central Asia in Kazakhstan, Kyrgyzstan and Tajikistan. Thrives in temperate climate, planted in well-drained soil and a position in full sun. **USES** The most popular fruit, grown and consumed in vast quantities worldwide. Eaten raw, cooked and in juice, and as apple cider, a light alcoholic drink. A reason for the fruit's popularity is its ability to be transported easily and stored for lengthy periods in cold stores. **ETYMOLOGY** Genus name is Latin for apple, and the epithet refers to the domestic cultivar of the species.

Almond ■ *Prunus dulcis*

DESCRIPTION Small deciduous tree with a compact crown. Grows to 6–8m in height. Leaves simple and lance shaped, with fine, serrated edges and arranged alternately. Flowers white with a pink centre, and comprise five petals. They bloom abundantly in March–April, and their fragrance attracts bees for pollination. Most parts of the plant are toxic. **HABITAT** Native to eastern Mediterranean region and south-west Asia. Cultivated in cooler climes in the Indian subcontinent, for example in Kashmir. Also adapts to hot, dry conditions of plains, as in Delhi, where it blooms profusely. **USES** Mainly cultivated commercially for the almond nut, known locally as *Badam*. Also used as an ornamental tree for its abundant blossoms, which cover the entire crown. Badamwari, Srinagar is a heritage garden that celebrates almond blossoms every spring. Almond milk and oil are derived from the nut. **ETYMOLOGY** Genus name is derived from the Greek *proume*, plum tree. The epithet *dulcis* means sweet.

Himalayan Cherry ■ *Prunus cerasoides*

DESCRIPTION Medium-sized deciduous tree with a characteristic flaky bark with a coppery shine. Grows to 8–10m in height. Leaves simple, and arranged alternately. They are oblong with acute pointed tips and fine, serrated edges. Flowers bloom profusely in hues of pink with dark pink centres, and are borne in clusters. They attract bees and other insects. Most part of the plant are toxic. **HABITAT** Native to east Asia in the Himalayan range, from Kashmir up to Myanmar. **USES** Ornamental cherry blossoms are celebrated around the world for their beauty. All parts of the plant are used in traditional medicine systems such as Ayurveda. Cherry brandy or liquor is made from the fruits. **ETYMOLOGY** Genus name is derived from the Greek *proume*, or plum tree, and the epithet *cerasoides* means like a cherry, from the Latin *cerasus*, or cherry.

Flowering Quince ■ *Chaenomeles japonica*

DESCRIPTION Deciduous, twiggy shrub with multiple branches and irregular form that grows to 2m tall. Oblong leaves simple, arranged alternately, and have fine, serrated

edges. Orange-red flowers attractive, cup shaped and borne along branches when nearly leafless. They bloom in April–May and attract bees for pollination, and are followed by small, apple-like fruits. **HABITAT** Native to Japan and South Korea. Cultivated in cooler areas of India like Kashmir and Kathmandu. A hardy plant, it also adapts to the heat in the plains, for example in Delhi. **USES** Ornamental shrub grown for its attractive flowers; also grown as a hedge. Fruits edible, raw or cooked, and used in traditional Kashmiri cuisine. **ETYMOLOGY** Genus name is derived from the Greek *chaino*, gape open, and *melon*, apple, referring to a mistaken belief that the apple-like fruits split open. The epithet *japonica* means from Japan.

Peach ■ *Prunus persica*

DESCRIPTION Small deciduous tree with a spreading crown and silvery bark. Grows to 5m in height. Lance-shaped leaves simple and arranged in opposite pairs. Light pink flowers have dark pink markings on petals and are borne directly on stems and branches. They bloom in March–April with the appearance of new leaves, and cover the entire crown with pink blossoms. Fruit is a fleshy drupe with a stone kernel covering the seed.
HABITAT Known to be a cultivated species, its earliest record of cultivation is in Xinjiang Province in north-west China. Widely cultivated in subtropical parts of India and Pakistan, and adapts to the hot, dry climate of Delhi. **USES** Popularly cultivated as a fruit tree, and also favoured for its spectacular blossoms.
ETYMOLOGY Genus name is derived from the Greek *proume* for plum tree. The epithet *persica* means of Persia, where peach trees were widely cultivated.

Hybrid Roses ∎ *Rosa x hybrida*

DESCRIPTION Semi-deciduous shrubs and climbers comprising miniature, shrub and climber varieties, growing to 0.5–6m in height. Leaves pinnately compound and arranged in opposite pairs. Leaflets oblong with pointed ends and fine, serrated edges. This group of modern rose cultivars includes Hybrid Tea roses, which are cultivars of European perpetual and Chinese tea roses. They are tall, hardy shrubs with an upright growth habit. Flowers are bud shaped with multiple petals, and borne solitarily on long stalks. They occur in a wide range of colours. Varieties such as 'Charles Mallerin' and 'Mirandy' are deep red in colour with a sweet fragrance. Floribunda roses are cultivars of Hybrid Tea and Polyantha roses. They bloom abundantly in clusters. Flowers are single or double with a flattish shape. The 'Else Poulsen' variety is one of the earliest Floribunda roses, with light pink flowers arranged in clusters. 'Iceberg' is another popular variety. Miniature roses are small shrubs with flowers that bloom abundantly singly or in clusters. Climber and rambler roses have a scandent habit and can be trained as climbers. **HABITAT** Hybrid cultivars are widely cultivated in gardens throughout the Indian subcontinent. They thrive in well-drained, porous soil that is acidic, and prefer a location in full sun. **USES** Ornamental shrubs grown for their highly attractive flowers. Entire gardens have been dedicated to the plant, such as the Rose Garden in the Chandigarh. Hybrid Tea roses are most popularly grown commercially for cut flowers. Essential oil from the petals, known as rose attar, is used in perfumes and cosmetics, and also as a food flavouring. Petals are used to make rosewater and a popular preserve known as *Gulkand*, as well as concentrates such as rose sherbets. The flowers are also used in traditional herbal remedies. **ETYMOLOGY** Genus name is derived from the Greek *rhodon*, rose.

Climber rose

Hybrid Tea roses 'Charles Mallerien', 'Doris Trysterman'

Floribunda roses 'Else Poulsen', Iceberg

Climbing Ylang Ylang ■ *Artabotrys hexapetalus*

DESCRIPTION Evergreen scrambling shrub with spreading habit. Grows to 3m in height. Glossy green leaves lance shaped and arranged in opposite pairs. Green flowers fade to yellow and camouflage well with the leaves, making them hard to detect. Intense fragrance of flowers in the evenings signals their presence. Flowers comprise six curved petals with

Ylang Ylang, Cananga odorata

a leathery texture, and are borne solitarily or in pairs; they bloom in summer. **HABITAT** Native to South India, Sri Lanka and south China. Cultivated in most parts of India, although not commonly seen, it thrives in hot, arid regions of North India as well as rainy, wet areas of north-west India. **USES** Cultivated for its fragrant flowers in gardens and parks. Floral oil derived from it is used in making perfumes. **ETYMOLOGY** Genus name is derived from the Greek *artao*, to support, and *botrys* refers to the bunches of grape-like fruits. The epithet *hexapetalus* refers to the six petals of the flower.

Kalanchoe ■ *Kalanchoe blossfeldiana*

DESCRIPTION Succulent perennial shrub with a compact form. Grows to 0.3–0.5m in height. Fleshy leaves simple and oblong with scalloped edges, and arranged in opposite pairs. Attractive red-pink flowers cruciform in shape, comprising four petals and arranged in large clusters. Cultivars such as *K. blossfeldiana* 'Calandiva' are double flowering, with more than 30 petals. They bloom in the cold months of winter to spring. **HABITAT** Native to Madagascar and cultivated widely in Indian gardens. Thrives in well-drained soil in a location in full sun, and also tolerates partial shade. **USES** Ornamental plant, commonly cultivated as a houseplant in pots. Can be grown indoors near windows that have good light. **ETYMOLOGY** Genus name is presumed to be derived from a Chinese vernacular name of a different species. The epithet *blossfeldiana* honours German horticulturist Robert Blossfeld.

Bird of Paradise ■ *Strelitzia reginae*

DESCRIPTION Evergreen shrub with clump-forming habit and leaves growing upright to a height of 1.5m. Greyish-green leaves paddle shaped with tapered ends and arranged spreading out like a fan. Showy, attractive flowers appear in summer up to early winter.

Blue-purple petals and orange sepals protected by boat-shaped spathe. **HABITAT** Native to South Africa. Widely cultivated in tropical and subtropical parts of the Indian subcontinent. Prefers a spot in a sunny position. **USES** Ornamental shrub grown for its attractive foliage and flowers. Flowers are long lasting and used in cut-flower arrangements. **ETYMOLOGY** Genus is named in honour of Queen Charlotte of the house of Mecklenburg-Strelitz, Germany. The epithet *reginae* means 'of the queen'.

Traveller's Palm
■ *Ravenala madagascariensis*

DESCRIPTION Handsome evergreen plant that is palm-like in form, with a single trunk fanning out into fronds, and a height of up to 8–9m. Leaves large, paddle shaped, similar to those of a banana, and spread out in one plane. Flowers dramatic, similar to those of *Strelitzia*, rising in sharp white petals from boat-shaped bracts that are tightly stacked. They bloom in summer, opening at night and attracting bats for pollination. In its native habitat, lemurs are known to pollinate the plant. **HABITAT** Native to Madagascar, and a plant of tropical climates. Cultivated widely in the Indian subcontinent's gardens, where it thrives in sunny locations. **USES** Grown as a specimen plant for its dramatic form and foliage. **ETYMOLOGY** Genus name is derived from vernacular Malagasy name *Ravinala*, which translates as leaves of the forests. The epithet name refers to its origins in Madagascar.

Lantana ■ *Lantana camara*

DESCRIPTION Vigorous evergreen shrub with a bushy, spreading habit and thorny branches. Grows to 2m in height. Aromatic leaves oblong with pointed ends and fine, serrated edges. Flowers minute and tubular, arranged in compact clusters. They are

multicoloured in combinations of pink and white or yellow and orange. They bloom nearly throughout the year, and attract birds, bees and butterflies for pollination. Colours darken after flowers are pollinated. **HABITAT** Native to Central and South America. Widely cultivated in the Indian subcontinent, where it has naturalized and grows wild along roadsides and wasteland. **USES** Hardy plant that requires little care, and is planted along roadsides and in parks. Unchecked, it can be an invasive plant. It can also be maintained as a hedge. Food plant for butterflies, and suited for growing in butterfly gardens. **ETYMOLOGY** Origin of the genus name is presumed to be derived from a Latin term for *Viburnum*. The epithet *camara* is a South American vernacular name for *Lantana*.

Trailing Lantana ■ *Lantana montevidensis*

DESCRIPTION Low-height, trailing shrub that covers the ground in thick foliage, and grows to 0.3m high. Leaves simple, arranged in opposite pairs, and have undersides that are covered in fine hairs. Mauve-coloured flowers bloom profusely in spring and sporadically throughout summer. **HABITAT** Native to tropical parts of South America, and widely cultivated in the Indian subcontinent. Prefers a position in full sun and also tolerates partial shade. **USES** Grown as a groundcover that forms a thick mat of foliage. A butterfly food plant and suited for growing in butterfly gardens. **ETYMOLOGY** Origin of the genus name is presumed to be derived from a Latin term for *Viburnum*. The epithet *montevidensis* refers to one of its places of origin in Montevideo, Uruguay.

Moss Verbena ▪ *Verbena tenuisecta* syn.*V. aristigera*

DESCRIPTION Perennial evergreen shrub that grows low and spreading, up to a height of 0.3m. Dark green leaves fine and deeply lobed, covering the ground like moss. They are arranged in opposite pairs. Purple or white flowers are profuse, small in size and arranged

in tight clusters. They bloom nearly all the year round and attract butterflies. **HABITAT** Native to South America from Brazil and Bolivia, to Argentina. Hardy plant cultivated widely throughout the Indian subcontinent. Thrives in full sun as well as in partial shade. **USES** Popularly used as a hardy groundcover. **ETYMOLOGY** Genus name is derived from the Latin for leafy twigs, which are used in rituals for making wreaths.

Petrea ■ *Petrea volubilis*

DESCRIPTION Large, woody climber with twining stems that grows to 6–8m in height. Leaves simple, elliptic in shape, and with a rough sandpaper-like texture. They are arranged in opposite pairs and in whorls. Lilac flowers cover crown in drooping spikes in spring, March–May, and sometimes also in autumn, October–November. They are star shaped with dark purple petals surrounded by lilac calyces that last much longer than the flower itself. White flower variant also occurs, though is not common. **HABITAT** Native to tropical America – Florida, Mexico and South America. Widely cultivated in gardens throughout the Indian subcontinent, and prefers a location in full sun. **USES** Grown as a garden ornamental for its abundant blossoms and characteristic twining stems. **ETYMOLOGY** Genus is named in honour of English patron of horticulture Lord Robert James Petre. The epithet *volubilis* describes the plant's twining habit.

Field Pansy ■ *Viola tricolor*

DESCRIPTION Annual herbaceous plant that grows to 0.2m tall. Light green leaves simple and arranged alternately. Attractive flowers small and borne solitarily, comprising

three colours – white, yellow and purple. They bloom in cold months in December–March and are lightly scented. Garden hybrids such as *V.* x *wittrokiana* have large blossoms in various shades of colour – white, pink, yellow, red and purple. **HABITAT** Native to Europe, ranging up to west Asia. Widely cultivated in the Indian subcontinent as a winter seasonal. Thrives in partial shade as well as full sun. **USES** Ornamental seasonal plant often used as a bedding or border plant in gardens. Flowers edible and can be used in salads or as garnishes. In Europe, they were traditionally used medicinally as an anti-inflammatory. **ETYMOLOGY** Genus name is derived from the Latin for sweet-smelling flowers. The epithet *tricolor* refers to the three colours of the flowers.

Pansy hybrid varieties

Blue Water Lily ■ *Nymphaea nauchali*

DESCRIPTION Perennial aquatic plant with rhizomes rooted in mud in shallow water. Round leaves have wavy margins and float on the surface of water. They have a waxy texture that repels water, and their undersides have deep, prominent veins. Mauve-blue flowers are fragrant, with a centre of yellow stamens tipped in mauve. They are borne solitarily on tall stalks that rise above the water's surface. They bloom in April–May and attract bees. **HABITAT** Native to tropical Asia, including India, Sri Lanka, Bangladesh and Myanmar. Grows at the edges of freshwater ponds and lakes. **USES** Ornamental marginal plant that is grown in shallow waterbodies for its attractive flowers. Flowers considered auspicious

and sold in markets as votive offerings. Rhizomes are medicinal, and are used in Ayurveda and Siddha systems; they are known to have astringent and antiseptic properties. **ETYMOLOGY** Genus name is derived from the Greek *Nymphe* for goddess of spring. The epithet *nouchali* is considered to be drawn from the Noakhali district in Bangladesh, where it is found in its natural habitat.

Water Lily ■ *Nymphaea pubescens*

DESCRIPTION Perennial aquatic plant with rhizomes rooted in mud in shallow water. Round leaves have wavy, toothed margins and float on the surface of water. They have a waxy texture that repels water, and their undersides are pink with a web of prominent veins. White or pink flowers open in the morning and close in the evening, and have a centre of yellow stamens. They are borne solitarily on tall stalks that rise above the water's surface. They bloom in summer to autumn, and attract bees. **HABITAT** Native to tropical Asia, including the Indian subcontinent. Grows in lakes and ponds across India, Sri

Lanka, Burma and Bangladesh. **USES** Ornamental marginal plant that is grown in shallow waterbodies for its attractive flowers. Flowers considered auspicious and sold in markets as votive offerings. Rhizomes and leaves used in traditional medicine remedies. **ETYMOLOGY** Genus name is derived from the Greek *Nymphe* for goddess of spring. The epithet *pubescens* refers to the fine hair covering on the undersides of the leaves.

Water Lily hybrid varieties

Mexican Sword Plant ■ *Echinodorous palifolius*

DESCRIPTION Perennial aquatic plant that grow from rhizomes submerged in shallow water and rises to 0.5m high above water. Leaves lance shaped with prominent

longitudinal veins, and upright in clumps. White flowers comprise three petals and yellow stamens in centre. They bloom sporadically throughout the year, and are arranged in whorls on a long, arching spike. **HABITAT** Native to tropical America from Mexico to Brazil, thriving in tropical climate along margins of waterbodies and streams. **USES** Grown as a marginal plant in aquatic gardens for its attractive foliage and flowers. Suited to growing in wetlands and reedbeds for treating grey water. **ETYMOLOGY** Genus name is derived from Greek and refers to the prickly fruit covering. The epithet *palifolius* means pale leaves.

Water Poppy ■ *Hydrocleys nymphoides*

DESCRIPTION Perennial aquatic plant that grows in shallow water, and rises to 0.2m in height. Leaves paddle shaped and rounded, floating or emerging from the water's surface. Flowers pale yellow with dark brown markings in centre, and cups shaped like poppies.

They bloom in summer and are borne solitarily. **HABITAT** Native to Central and South America, occurring in swamps and wetlands in its natural habitat. Thrives in tropical climate and cultivated in aquatic gardens in the Indian subcontinent. **USES** Cultivated in gardens for its pretty poppy-like flowers, and often grown in large pots filled with water. **ETYMOLOGY** Genus name is derived from the Greek *hydor*, water, and *kleis*, key. The epithet *nymphoides* refers to its resemblance to the water lily or Nymphaea family.

Water Hyacinth ▪ *Eichhornia crassipes*

DESCRIPTION Perennial aquatic plant that floats on the water's surface and can grow to 0.3m in height. Glossy green leaves simple, spongy and swollen, with rounded ends. Purple flowers are showy, and borne in clusters on an erect spike. Central petal has a marking in dark purple and yellow that acts as a guide for insect pollinators. Flowers bloom in August–September and open at night. **HABITAT** Native to Brazil, and has naturalized in the Indian subcontinent. Grows wild in freshwater lakes and rivers. **USES** Planted in bog and water gardens for its attractive flowers. Popularly used in waste-water treatment, since it removes heavy metals and other pollutants from the water. An invasive plant, its growth should be kept in check. **ETYMOLOGY** Genus is named in honour of a Prussian minister of education, J. A. Fr. Eichhorn. The epithet *crassipes* means thick footed, referring to the swollen, bulbous stems.

Fried Egg Tree ■ *Oncoba spinosa*

DESCRIPTION Small evergreen tree with a spreading crown. Grows to 4–5m in height. Oblong leaves have tapered ends and fine, serrated edges. They are simple and arranged alternately. Flowers attractive and fragrant, comprising five white petals surrounding a yellow centre of stamens. Flower composition resembles a fried egg, hence the common name. Flowers bloom in spring, in March–May, and are borne solitarily. **HABITAT** Native to Arabia and tropical parts of Africa. Cultivated in India across a large geographical region from Tamil Nadu, Karnataka and Maharashtra, to Uttar Pradesh and Delhi. **USES** Cultivated in gardens for its showy, fragrant blossoms. Snuff boxes are made from the hard shell of its fruits in Africa, hence it is also known as the Snuff Box Tree. **ETYMOLOGY** Genus name is known to be derived from the Arabic name *Onkub* for a species in North Africa. The epithet *spinosa* refers to spiny, or thorns at the bases of the leaves.

Purple Shamrock ■ *Oxalis triangularis*

DESCRIPTION Perennial herbaceous shrub with clump-forming habit. Grows from tuberous roots to 0.3m in height. Purple leaves are trifoliate and borne on slender stems, and leaflets are triangular in shape. Flowers lilac and funnel shaped. They appear nearly throughout the year in small clusters, and attract insect pollinators. A yellow-flowering species, *O. stricta*, is popularly grown as a winter seasonal and has bright green leaves.
HABITAT Native to South America – Peru, Brazil and Argentina. Not common in Indian gardens but becoming increasingly popular. Prefers moist soil conditions and a location in partial shade. **USES**

Ornamental groundcover plant grown for its attractive foliage and flowers. Also suited for growing as an indoor pot plant in colder climates. Leaves and flowers edible and used as a garnish or in salads.
ETYMOLOGY Genus name is derived from the Greek *oxalis*, acid, since the plant contains oxalic acid, which imparts a sour taste to the flowers and leaves. The epithet *triangularis* refers to the triangular leaflets.

Further Reading

BNHS, Kehimkar Isaac. 2001. *Common Indian Wild Flowers*. Oxford.

Bor, N. L. & Raizada, M. B. 1982. *Some Beautiful Indian Climbers & Shrubs*. Bombay Natural History Society.

Buchmann, Stephen. 2015. *The Reason for Flowers*. Simon & Schuster.

Fayaz, Ahmed. 2011. *Encyclopedia of Tropical Plants*. UNSW Press.

Harrison, Lorraine. 2018. *Latin for Gardeners*. Mitchell Beazley.

Hodge, Geof. 2013. *RHS Botany for Gardeners*. Mitchell Beazley.

Howell, Catherine H. 2009. *Flora Mirabilis*. National Geographic. Timber Press, Portland, Oregon, USA.

Krishen, Pradip. 2006. *Trees of Delhi*. Penguin India.

Llamas Kirsten Albrecht. 2003. *Tropical Flowering Plants, A Guide to Identification and Cultivation*.

Patnaik, Naveen. 1994. *The Garden of Life*. Aquarian Press.

Pavord, Anna. 2007. *The Naming of Names – The Search for Order in the World of Plants*. Bloomsbury Press.

Pell, Susan K. & Bobbi, Angell. 2016. *A Botanist's Vocabulary*. Timber Press, Portland, Oregon, USA.

Randhawa, G. S., Mukhopadyay, A. N. & Mukhopadyay, A. 1998. *Floriculture in India*. Allied Publishers Pvt. Ltd.

Riffle, Robert Lee. 1998. *The Tropical Look*. Thames & Hudson.

Sahni, K. C. 1998. *The Book of Indian Trees*. Oxford.

Yong, Jean W.H., Tan Puay Yok, Nor Hafiz Hassan & Tan Swee Ngin. 2010. *A Selection of Plants*. National Parks Board, Singapore Botanical Garden.

Websites

www.missouribotanicalgarden.org
www.plantsoftheworldonline.org
www.theplantlist.org
www.florafaunaweb.nparks.gov.sg
www.nybg.org/science-project/world-flora-online
www.flowersofindia.net

ACKNOWLEDGEMENTS

The authors would like to thank their colleagues at PSDA Studio New Delhi for supporting us while this book was being researched. In particular Madhu Shankar, Vishwesh Vishwanathan, Arti Mathur. Vaibhav Aggarwal made the biodiversity regions map. Oken Tayeng sent pictures of rhododendrons from Arunachal Pradesh. Gautam Sachdeva along with Divya Saraf took pictures at Agro Horticultural Society Kolkata, Arti Mathur took some pictures in our studio garden.
TR Ramakrishnan, "Thambi" gave valuable inputs in editing the text
As usual, gratitude to Ven. Nicholas Vreeland for tips on photography.

▪ INDEX ▪